SOLID STATE PHASE TRANSFORMATIONS

V. RAGHAVAN
Formerly Professor of Materials Science
Indian Institute of Technology, Delhi

PHI Learning Private Limited
Delhi-110092
2025

In fond memory of ***Shri Asoke K. Ghosh*** *(October 1942 – February 2024), Founder Chairman and Managing Director of PHI Learning, whose vision endlessly inspires.*

The Legacy Continues....

Published by Pushpita Ghosh, PHI Learning Private Limited, Rimjhim House, 111, Patparganj Industrial Estate, Delhi-110092 and Printed by Syndicate Binders, A-20, Hosiery Complex, Noida, Phase-II Extension, Noida-201305 (N.C.R. Delhi).

₹525.00

SOLID STATE PHASE TRANSFORMATIONS
by V. Raghavan

ISBN-978-81-203-0460-4 (Print Book)
ISBN-978-93-90669-76-9 (e-Book)

Foreword

The interrelationships between structure and properties constitute a central theme in physical metallurgy and materials science, and phase transformations offers one of the major methodologies through which structures can be changed and controlled. Accordingly, the subject of phase transformations now occupies a core position in the teaching of physical metallurgy and materials science. Indeed, the present text was written in recognition of that fact, i.e., to provide a teachable balance between the advanced concepts reached in this field and the constraints of covering the subject matter adequately in a single course.

Due attention is given to the main fundamental aspects of phase transformations, such as diffusivity, thermodynamic driving forces, nucleation and growth kinetics, diffusional and diffusionless phase changes, and resulting morphologies. Also included are separate chapters on dispersed-phase coarsening and recrystallization, both of which are important types of structural change and depend on phenomena characteristic of phase transformations.

The author's long experience with phase transformations, in research as well as in teaching, has enabled him to cover the relevant topics in a lucid and succinct manner. Moreover, even though much of the underlying knowledge has evolved from metallic systems, the treatment is presented in a way that is generally applicable to nonmetallic materials. The overall field of phase transformations remains both challenging in theoretical advances and vital in the development of high-performance materials. International conferences on diverse manifestation of the subject continue to proliferate, leading to valuable treatises. But the book at hand fills a different need, namely as a text designed specifically for teaching purposes.

Cambridge, MA
USA.

MORRIS COHEN

Preface

This book is the outcome of the author's experience spread over a period of 20 years in teaching a course on Solid State Phase Transformations to Metallurgy and Materials Science majors and to junior postgraduate students. The material was first put together in the form of lecture notes, when the author was teaching the course in association with Professor Morris Cohen at the Massachusetts Institute of Technology in the years 1969-71. The notes have since undergone updatings as found necessary.

The opening chapter is a brief introduction to the various types of phase transformations. In the second chapter, the principles of diffusion are dealt with, as many transformations occur through diffusional processes. The principles of the nucleation and growth kinetics are discussed in some detail in two separate chapters. The applications include continuous precipitation, particle coarsening, the pearlitic reaction and the martensitic transformations. The last chapter discusses spinodal decomposition as an example of a homogeneous transformation. The emphasis is throughout on basic principles. Where controversial issues are involved, the author has tried to present what, in his opinion, is the generally-accepted view. The student who chooses to do research in this area will surely acquaint himself with the "full truth" later.

Suggestions for further reading and some exercises are given at the end of each chapter. The meaning of symbols is explained in the text, when they are introduced first. In addition, a list of symbols is included at the end of the book for ready reference.

Grateful thanks are due to Professor Morris Cohen for making it possible for the author to teach this course at MIT for 2 years. When it became clear that he is not going to be involved in the book-writing, Professor Cohen most graciously agreed to the author doing it on his own, using the jointly-prepared lecture notes. Thanks are due to the Indian Institute of Technology, Delhi for partial financial assistance in preparing the manuscript and the diagrams.

V. RAGHAVAN

New Delhi,
July 1986.

Contents

1

Introduction

Understanding the relationship between the microstructure and the properties of materials is the main concern of metallurgists, ceramists and other materials scientists. The aim of a heat treatment or a thermomechanical treatment is to obtain a microstructure that yields the desired properties. During the treatment, a particular phase transformation might have to be induced or suppressed for this purpose. In order to do this effectively, it is necessary to understand all aspects of the corresponding phase transformation. In this book, we deal with the fundamental principles that govern the kinetics and the mechanism of solid state phase transformations.

1.1 DEFINITION OF A PHASE CHANGE

In the classical definition, a phase is a physically distinct, chemically homogeneous and mechanically separable part of a system. Two phases are distinguishable from each other, if they form different states of aggregation (solid, liquid and vapour) or in the same state of aggregation, if they have different compositions or different crystal structures. For the same composition and crystal structure, differences in the electronic structure are also sometimes used to distinguish different phases.

A phase transformation or a phase transition is defined as the change from one or more phases (called the parent phases) to one or more other phases (called the product phases). It follows then, from the above definition of a phase, that a phase transformation involves changes in the

1 state of aggregation,
2 composition,
3 crystal structure, or
4 electronic structure.

It may be also a combination of more than one of the above changes.

The terms 'transformation', 'transition' and 'reaction' are often used interchangeably. The first term appears to have wider usage, whereas the term 'transition' generally refers to thermodynamically higher order changes such as electronic transitions.

1.2 ATOM MOVEMENTS IN PHASE TRANSFORMATIONS

As we are concerned with solid state phase transformations, we shall not consider changes in the state of aggregation while discussing the atom movements involved in transformations.

No atom movements take place during changes in the electronic structure. Changes in composition and crystal structure in the solid state clearly require the movement of atoms within the solid. The nature and the extent of such atom movements differ widely in different transformations and fall under the following three categories:

1 movements over a large number of interatomic distances;
2 movements over one or two interatomic distances; and
3 movements over a fraction of an interatomic distance.

Long-range and Short-range Diffusion

The atom movements of the first two categories above are brought about by the process of diffusion in the solid state. Diffusion is the mass flow process by which atoms change their positions relative to their neighbours by "random-walk". The phenomenon of diffusion in the solid state is discussed in Chapter 2. Changes in composition require atom movements over a large number of interatomic distances and the corresponding process is called *long-range diffusion.*

When a change in crystal structure takes place, the atom movements may be only over one or two interatomic distances. The atoms may be simply transferred from the parent crystal structure to the product structure across an interface which is one or two interatomic distances in thickness. This process is called *short-range diffusion.*

Diffusionless Changes

In category 3 above, the atoms may move only through a fraction of an interatomic distance. Such movements bring about crystal structure changes. The product crystal structure can be generated here only when the atom movements occur in a *coordinated fashion.* Otherwise, an amorphous product will result. The random-walk diffusion referred to earlier can be compared to the way civilians perform a task each in his own way, while the coordinated movement of atoms has been called a military transformation by Christian. In the absence of interchange of atom positions by random walk, the military transformations are said to be diffusionless.

1.3 TYPES OF PHASE TRANSFORMATIONS

Transformations with a Change in Composition

Consider the miscibility gap in the Al—Zn phase diagram, illustrated in Fig. 1.1. Above 352°C, an Al-39.5 at. % Zn alloy is a single phase, α, with the FCC crystal structure. On cooling below 352°C, say, to 300°C, this

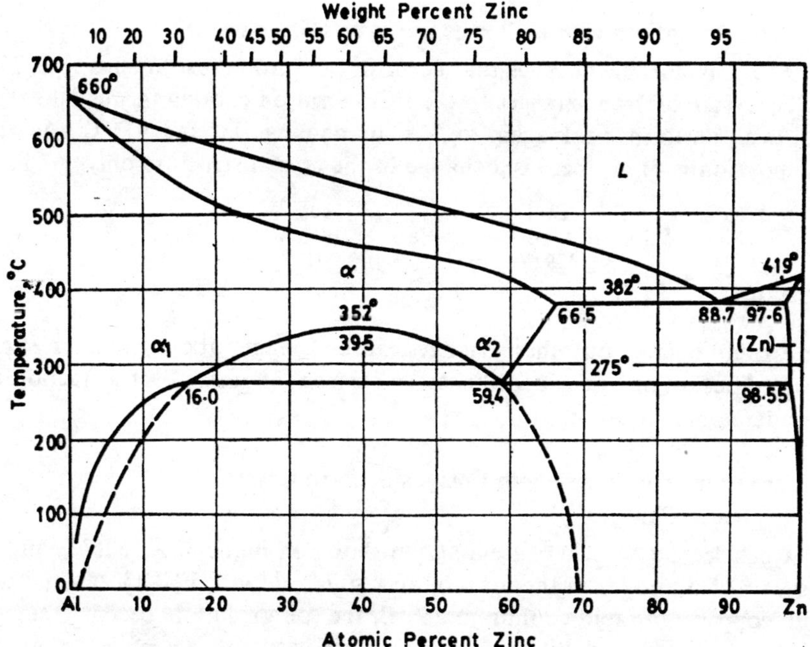

Fig. 1.1 The Al–Zn phase diagram depicting the miscibility gap in the α region. The dotted lines are the metastable extensions of the gap to lower temperatures.

single phase separates into two phases α_1 and α_2 with different zinc contents. There is no change in crystal structure.

$$\alpha \rightarrow \alpha_1 + \alpha_2$$

Structure:	FCC	FCC	FCC
Zn, at. %:	39.5	21	56

A transforming region of the α phase will be split into two regions α_1 and α_2, after the transformation as shown schematically in Fig. 1.2. Clearly, this requires that the atoms diffuse at least through a distance comparable to the smallest dimension of the transforming region. A typical size of such a transforming region in the solid state is 1 μm, which means that diffusion has to occur over several thousands of interatomic distances to bring about the compositional change. Thus, long range diffusion is required here.

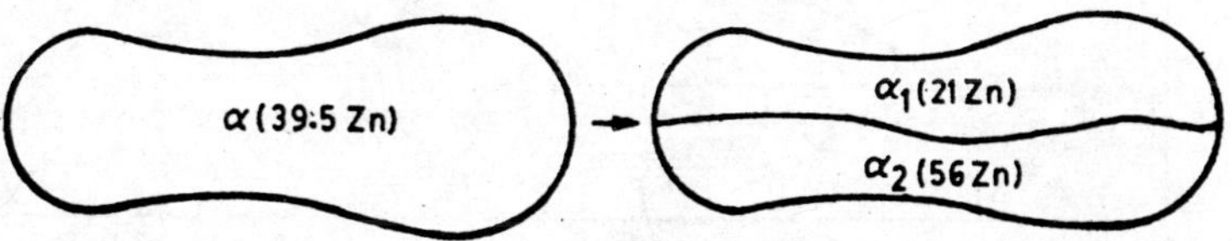

Fig. 1.2 Schematic illustration of a transformation with a composition change.

Transformations with a Change in Crystal Structure

The polymorphic changes that occur in pure iron illustrate this type. The BCC crystal of iron changes to the FCC form on cooling through 1392°C and this transforms back again to BCC on cooling through 911°C. As the composition is fixed, there is a change in the crystal structure only.

$$\delta \text{ Fe} \xrightarrow[1392^\circ\text{C}]{\text{cool}} \gamma \text{ Fe} \xrightarrow[911^\circ\text{C}]{\text{cool}} \alpha \text{ Fe}$$

Structure: BCC FCC BCC

As already pointed out, these changes can be brought about by either short range diffusion or by the coordinated movement of atoms over a fraction of an interatomic distance.

Transformations with both Composition and Crystal Structure Changes

Consider the Al—Cu phase diagram shown in Fig. 1.3. At 525°C, an Al-4.5 wt.% Cu alloy is in the form of an α solid solution (FCC) just saturated with copper. On rapid cooling to 250°C, the solid solution becomes supersaturated, as the equilibrium solubility of copper decreases to about

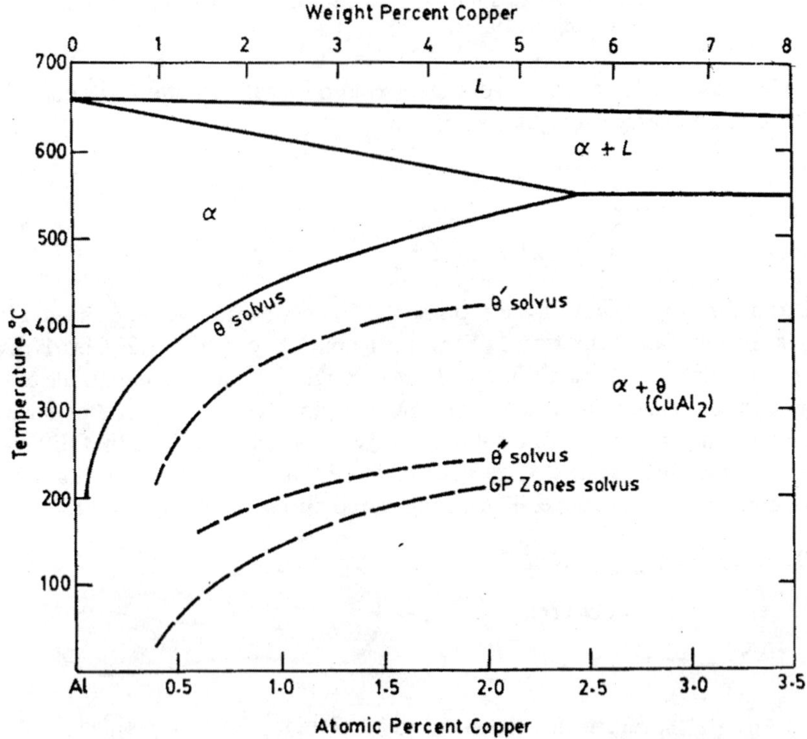

Fig. 1.3 The aluminium-rich end of the Al–Cu phase diagram.

0.3 wt.%. The excess copper precipitates from the matrix in the form of θ ($CuAl_2$) particles with the BCT structure. This process is called *continuous precipitation* and the reaction can be written as:

	$\alpha_{supersat}$	$\rightarrow$	α_{sat}	+	θ ($CuAl_2$)
Structure:	FCC		FCC		BCT
Cu, wt.%:	4.5		0.3		52

The stable precipitate θ may not always form. The kinetics of precipitation may be greatly facilitated by the formation of metastable transition precipitates: GP zones, θ'', or θ'. The metastable solvus boundaries corresponding to these precipitates are shown in Fig. 1.3. Note that the solubility of copper increases, as the precipitate becomes increasingly metastable in the order: $\theta \rightarrow \theta' \rightarrow \theta'' \rightarrow$ GP zones.

Austenite of 0.8% C decomposes on cooling through 727°C to pearlite, a mixture of BCC ferrite of 0.02% C and orthorhombic cementite of 6.69% C.

			Pearlite		
	Austenite	$\xrightarrow{\text{cool}}$	Ferrite	+	Cementite
Structure:	FCC		BCC		Orthorhombic
C, wt.%:	0.8		0.02		6.69

This transformation called the *discontinuous precipitation* is discussed in Chapter 8. A plate-like region of austenite transforms to a pair of parallel plate-like ferrite and cementite crystals. The compositional changes are brought about by the long-range diffusion of the interstitial carbon. The diffusion distance is of the order of the thickness of the transforming plate-like region.

Substitutional diffusion is involved in the transformation of an Fe-7 at.% Ni alloy on cooling from 750°C to 650°C, when it decomposes to two different phases: BCC α (5 at. % Ni) and FCC γ (13 at. % Ni), see the Fe—Ni phase diagram in Fig. 1.4.

Substitutional diffusion is very slow as compared to interstitial diffusion; so the above phase change rarely occurs at ordinary cooling rates. Instead, the 7 at.% Ni alloy on further cooling below 600°C may simply transform from the FCC to the BCC structure without any change in composition. If such a change occurs by means of short-range diffusion, it is called a *massive transformation* (Chapter 9). If the change occurs by the coordinated movements of atoms, it is called a *martensitic transformation* (Chapter 11). The 0.8% C steel in the previous example might also transform to martensite in a similar fashion on rapid quenching:

	austenite	$\xrightarrow{\text{quench}}$	martensite
Structure:	FCC		BCT
C, wt.%:	0.8		0.8

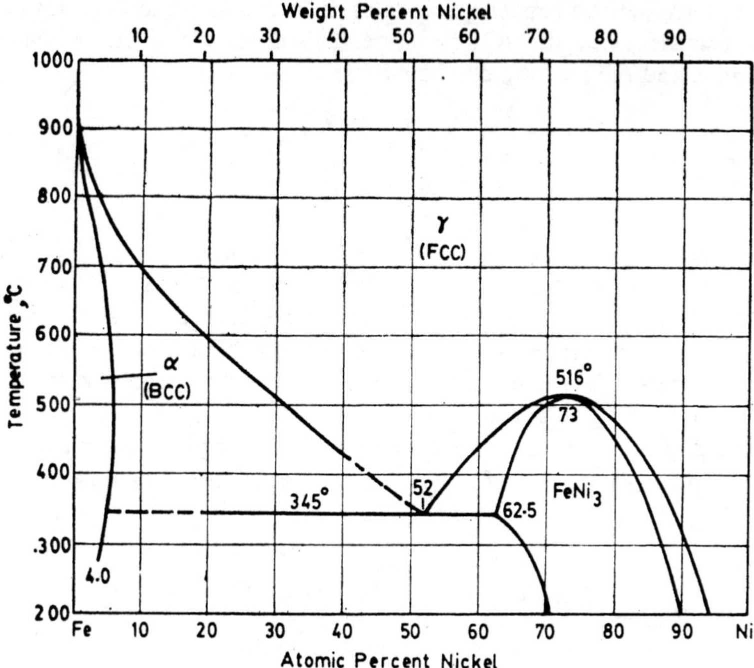

Fig. 1.4 The Fe–Ni phase diagram between 1000 to 200°C.

Here, the cooling rate is fast enough to prevent even the interstitial diffusion of carbon. In alloy steels, it is quite possible to have an intermediate situation, where the compositional changes corresponding to interstitial diffusion occur but not those corresponding to substitutional diffusion of alloying elements such as Ni, Cr and Mn.

Transformations with a Change in Order

A disordered solid solution can transform to an ordered solid solution on cooling through a critical temperature T_c, as schematically shown in Fig. 1.5. A Cu-50 at.% Zn alloy is a single β phase and is in the disordered state above 479°C. The copper and zinc atoms are randomly distributed

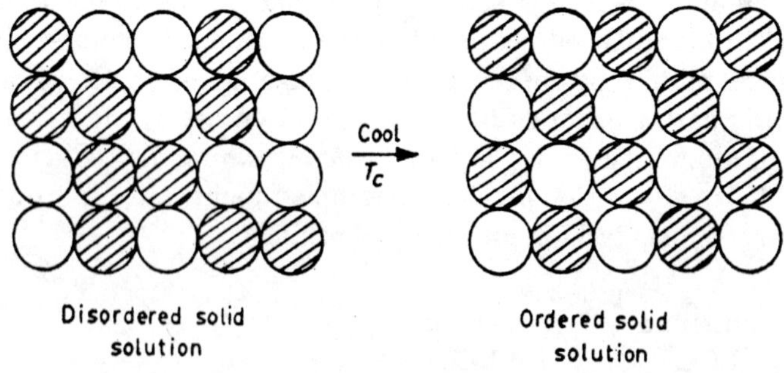

Fig. 1.5 The disordered and ordered states of a solid solution.

on the BCC sites. On cooling through 479°C, order sets in and the copper atoms occupy the body corners and the zinc atoms occupy the body centres (Note that the terms body corner and body centre are interchangeable in the BCC crystal). On ordering, the space lattice of the structure becomes simple cubic and additional new lines appear on an x-ray diffraction pattern. The ordered structure is called a *superlattice* or a superstructure. As the copper and zinc atoms have to switch places appropriately within the crystal to bring about order, short range diffusion is required.

Electronic Transitions

The transition from the paramagnetic to the ferromagnetic state on cooling iron through the Curie temperature of 769°C serves as an example of this type. The electron spins which are randomly oriented in the paramagnetic state align themselves in a parallel fashion on cooling through the Curie temperature T_c.

	Fe	$\xrightarrow[769°C]{cool}$	Fe
electron	paramagnetic		ferromagnetic
spin direction:	↓↓↑↓↑↑↑↓↓↑↑↑↓		↑↑↑↑↑↑↑↑↑↑↑↑↑

A similar transition occurs, when paramagnetic manganese oxide is cooled through the Neel temperature of −157°C, the product phase being the antiferromagnetic MnO:

	MnO	$\xrightarrow[-157°C]{cool}$	MnO
electron	paramagnetic		antiferromagnetic
spin direction:	↓↓↓↑↑↓↑↑↑↓↓↓↑		↑↓↑↓↑↓↑↓↑↓↑↓↑

No atom movements take place during these electronic transitions. The realignment of the electron spins occurs very rapidly, so that the equilibrium state is attained as soon as the crystal attains the temperature corresponding to that equilibrium state.

1.4 HOMOGENEOUS VERSUS HETEROGENEOUS TRANSFORMATIONS

A transformation that takes place more or less simultaneously in all parts of an assembly is regarded as a homogeneous transformation. Reactions in the gaseous phase are homogeneous. Some liquid and solid-state transformations are also homogeneous. Changes involving electronic transitions are homogeneous, taking place simultaneously throughout the system. An interesting case of a homogeneous transformation in the solid-state is the *spinodal decomposition*, discussed in Chap. 12. In this case, the transformation starts as a small composition fluctuation spread over a large volume of the material. Initially, there is no sharp boundary between the parent and the product phases. The compositional fluctuation grows in intensity with time to finally yield the equilibrium phases. This is illustrated in Fig. 1.6(a). The transformation in an Al-39.5 at. % Zn alloy discussed earlier may occur as a homogeneous transformation, the compositional fluctuations being brought about by long range diffusion.

A heterogeneous transformation is of the *nucleation-and-growth* type. Tiny volumes of the product phase called nuclei, often assumed to be the same in structure and composition as the transformation product, form first. A sharp boundary delineates the nuclei from the surrounding matrix. These small regions subsequently grow by the outward movement of the boundary, with corresponding changes in composition (and crystal structure) behind the advancing front. This is illustrated in Fig. 1.6(b). The long range diffusion

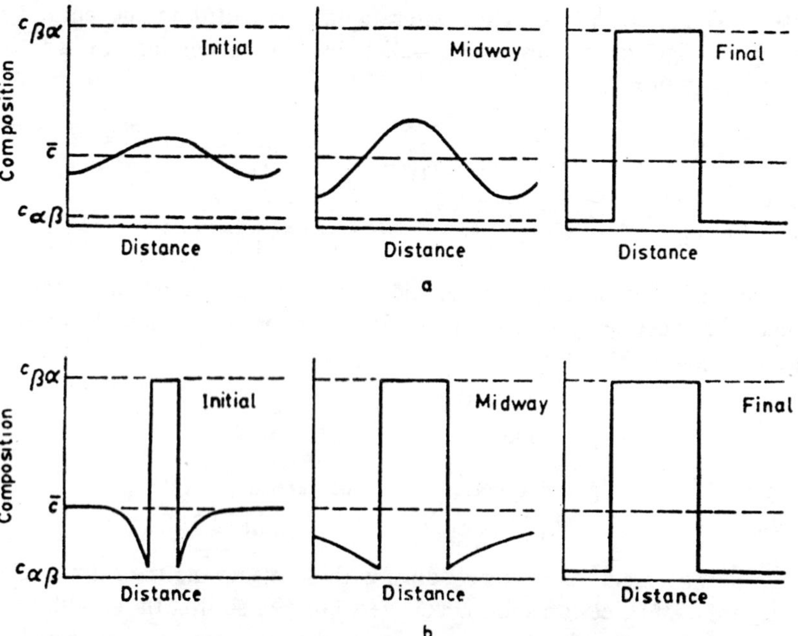

Fig. 1.6 (a) Transformations occurring by compositional fluctuations, and (b) transformations occurring by nucleation and growth.

necessary to bring about the compositional changes occurs in the surrounding matrix. The principles of nucleation and growth are discussed in Chaps. 4 and 5. A large number of transformations discussed in this book are of the nucleation-and-growth type.

1.5 BUERGER'S CLASSIFICATION

Buerger has given a structural classification of solid-state transformations based on changes in coordination and bond type. His categories are:

1 Transformations of secondary coordination

Displacive	rapid
Reconstructive	sluggish

2 Transformations of order-disorder
   Rotational — rapid
   Substitutional — sluggish

3 Transformations of first-coordination
   Dilatational — rapid
   Reconstructive — sluggish

4 Transformations of bond type (usually sluggish)

Transformations of secondary coordination (category 1) occur in network structures having atoms of low primary coordination. For example, in silicate structures, a transformation of secondary coordination takes place by rearrangement of the silicate tetrahedra relative to one another but without change in the primary coordination within the tetrahedra. This transformation can be brought about in two ways: (a) by displacing the tetrahedra with respect to one another without breaking the bonds between them, i.e., as a diffusionless, displacive transformation, and (b) by breaking the bonds between the tetrahedra and then rearranging them in a new array corresponding to the product crystal structure, i.e., a reconstructive transformation brought about by short-range diffusion.

Category 2, transformations of order-disorder, have also two subsets. Groups of tightly bound atoms in an ordered structure can rotate relative to the rest of the structure and so induce disorder. Likewise, interchanging positions among atoms in a random fashion can cause disordering. These two possibilities correspond to the rotational and substitutional types in Buerger's classification.

As in category 1, transformations of first coordination can also occur with or without breaking bonds. These two possibilities are designated as reconstructive and dilatational types in category 3 and correspond to transformations that occur by means of short-range diffusion and those that are diffusionless. No transformations requiring long range diffusion are included in Buerger's classification.

FURTHER READING

J.W. Christian, *The Theory of Transformations in Metals and Alloys*, General Introduction, Chapter 1, p. 1, Pergamon Press, Oxford (1975).

M.J. Buerger, *Phase Transformations in Solids*, p. 183, John Wiley, New York (1951).

EXERCISES

1.1 Give two examples of phase transformations where the proper control of the transformation can result in different combinations of properties.

1.2 List the possible differences in the nature of short-range diffusion in a polymorphic transformation versus an order-disorder transformation.

1.3 What types of phase changes require long-range diffusion?

1.4 In what type of transformation is a composition change not possible?

1.5 Can a ferromagnetic to paramagnetic change be described as an order-disorder transformation? Explain.

2

Diffusion in Solids

A number of phase transformations in solids are brought about by the process of diffusion. The study of diffusion is vital to the understanding of the structure-property relationships in materials, as structural control to achieve the optimum properties is often dependent on the rate of diffusion.

In this chapter, we first consider the macroscopic laws of diffusion. The atomistic model of diffusion is discussed in the later sections.

Diffusion is defined as the mass flow process by which atoms or molecules change their positions relative to their neighbours within a phase under the influence of thermal energy and a gradient. This gradient may be of chemical potential resulting from a concentration gradient, or a gradient due to temperature, stress, electric field or gravitational field. We will deal with concentration gradients only.

2.1 FICK'S LAWS OF DIFFUSION

The concentration gradients are used in laws first proposed by Adolf Fick in 1855. For ideal solutions, these gradients are directly related to the chemical potential gradients. Here, only unidirectional diffusion is considered.

Fick's First Law

The first law of Fick states

$$\frac{\mathrm{d}n}{\mathrm{d}t} = -DA\frac{\mathrm{d}c}{\mathrm{d}x} \tag{2.1}$$

where

$\frac{\mathrm{d}n}{\mathrm{d}t}$ is the number of moles of the diffusion specie passing through a cross-sectional area A perpendicular to the diffusion direction x per unit time,

c is the concentration of the specie in mol m^{-3}, and

D is the proportionality factor known as the *diffusivity or diffusion coefficient*.

The negative sign on the right side of Eq. 2.1 indicates that the mass flow is *down* the concentration gradient. D is a material property that depends on the diffusing specie, the composition of the medium in which diffusion occurs and the temperature. It is easy to show that its units are $m^2 s^{-1}$.

The first law can also be expressed in terms of *flux* J:

$$J = -\frac{1}{A}\frac{dn}{dt} = -D\frac{dc}{dx} \tag{2.2}$$

This form of Fick's law is identical to that of Fourier's law for heat flow or Ohm's law for electrical charge flow.

Steady state flow is obtained when the flux at any cross-sectional plane along the diffusion distance remains constant. Here, the concentration-distance profile does not vary with time. If the diffusivity D is a function of concentration, for a concentration c' at distance x' and c'' at distance x'', we have the condition for a constant flux:

$$D_{c'}\left(\frac{dc}{dx}\right)_{c'} = D_{c''}\left(\frac{dc}{dx}\right)_{c''} \tag{2.3}$$

When $D_{c'} \neq D_{c''}$, it follows that $(dc/dx)_{c'} \neq (dc/dx)_{c''}$. In other words, the concentration-distance profile is not a straight line, Fig. 2.1. If D is independent of composition, the gradient is constant and the profile is a straight line, see Fig. 2.1.

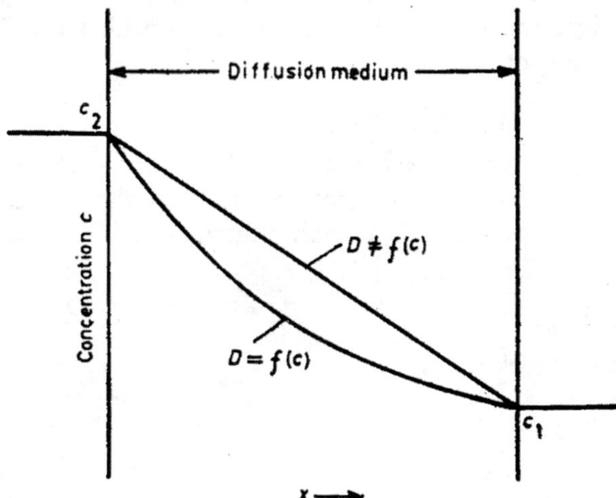

Fig. 2.1 Concentration-distance profile for steady state flow.

Fick's Second Law

Fick's second law corresponds to *non-steady state flow*. Here, at a given instant, the fluxes at different cross-sectional planes along the diffusion direction are not equal. Also, the flux at a given cross-section is a function of time. Consequently, the concentration-distance profile changes with time.

This is the situation most frequently met in applications of the diffusion laws.

Consider an elemental volume of length Δx along the diffusion distance x and of unit cross-sectional area (perpendicular to the diffusion distance), Fig. 2.2. The volume of such an element is Δx. The rate of accumulation (or

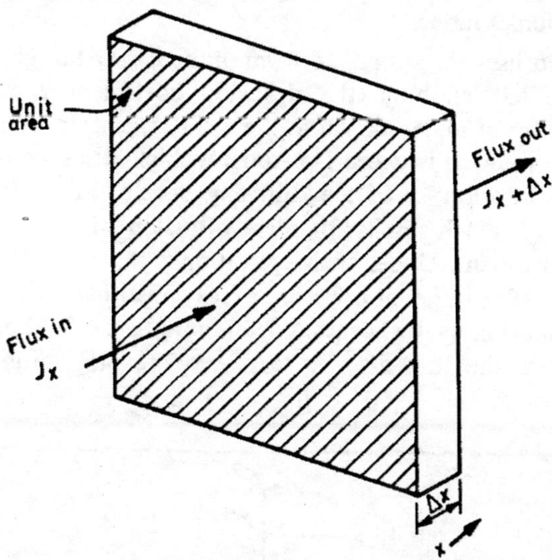

Fig. 2.2 In nonsteady state flow, the fluxes in and out of an elemental volume are not equal.

depletion) of the diffusing specie within this elemental volume is $(\partial c/\partial t)\, \Delta x$ and can be expressed in terms of fluxes into and out of the volume:

$$\frac{\partial c}{\partial t}\,\Delta x = J_x - J_{x+\Delta x} \tag{2.4}$$

Substituting $J_{x+\Delta x} = J_x + (\partial J/\partial x)\, \Delta x$ into Eq. 2.4, we get

$$\frac{\partial c}{\partial t} = -\frac{\partial J}{\partial x} \tag{2.5}$$

Using Eq. 2.2 in 2.5,

$$\frac{\partial c}{\partial t} = -\frac{\partial}{\partial x}\left(-D\,\frac{\partial c}{\partial x}\right)$$

$$= \frac{\partial}{\partial x}\left(D\,\frac{\partial c}{\partial x}\right) \tag{2.6}$$

Eq. 2.6 is Fick's second law for unidirectional flow under nonsteady state conditions.

If D is independent of concentration, Eq. 2.6 simplifies to

$$\frac{\partial c}{\partial t} = D\,\frac{\partial^2 c}{\partial x^2} \tag{2.7}$$

Even though D frequently varies with concentration, solutions of the differential Eq. 2.7 are quite commonly used in practice, because of the relative simplicity of the solution.

2.2 SOLUTIONS TO FICK'S SECOND LAW

The Thin Film Solution

This solution uses the simplified form of the second law given by Eq. 2.7. It can be used to determine the diffusion coefficient D experimentally. A thin disc or film of material rich in the diffusing specie (concentration c_2) is welded as a sandwich between two rods or thick discs containing a low (or zero) concentration c_1 of the diffusing specie. The thickness a of the sandwiched film is much smaller than the diffusion distance to be encountered in the experiment. Under such conditions, the concentration-distance curves obtained after different diffusion times constitute a set of symmetrical Gaussian distribution curves, centred at the middle of the thin disc, Fig. 2.3. The midpoint of the thin disc is taken as the origin for specifying the diffusion distance x.

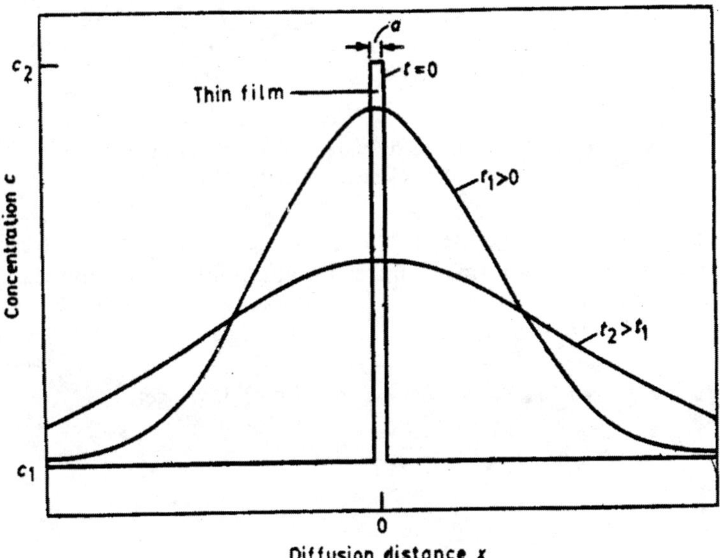

Fig. 2.3 Concentration-distance profiles in the thin film solution.

The corresponding solution to Eq. 2.7 is

$$c(x, t) - c_1 = \frac{\alpha}{2\sqrt{\pi D t}} \exp\left(-\frac{x^2}{4Dt}\right) \tag{2.8}$$

where

$$\alpha = (c_2 - c_1)\, a$$

$$= \int_{-\infty}^{+\infty} (c - c_1)\, dx \tag{2.9}$$

By differentiating Eq. 2.8, it is seen that Eq. 2.7 is obtained, showing that Eq. 2.8 is one of the possible solutions.

The initial conditions of this set up are

1 $t = 0, \quad x = 0, \quad c = \infty$; and

2 $t = 0, \quad x \neq 0, \quad c = c_1$.

Condition 1 results, because it is assumed that all the diffusing specie in the initial thin film is concentrated on a plane of zero thickness. The boundary conditions are

3 $t > 0, \quad x = \pm\infty, \quad c = c_1$; and

4 $t = \infty$, for all x, $\quad c = \bar{c} \approx c_1$.

Condition 4 results from the fact that the volume of the thin film is negligible compared to the volume of the entire set-up.

After the diffusion anneal, the concentration c can be determined as a function of the diffusion distance x, by means of careful sectioning and chemical analysis. The experimental data are plotted as log $(c - c_1)$ versus x^2. This should yield a straight line, from the slope of which D can be determined using Eq. 2.8.

The diffusion of atoms in a medium consisting of the same atoms is called *self-diffusion*. When this is to be determined, radioactive isotopes are used in the thin film. After the diffusion anneal, the radioactivity is determined as a function of the diffusion distance.

The Grube Solution

The experimental set-up here consists of a diffusion couple, i.e., two long bars welded face-to-face, the concentration of the diffusing specie in one, c_2, being higher than that in the other, c_1. The concentration-distance profiles after different lengths of diffusion annealing time are as shown in Fig. 2.4.

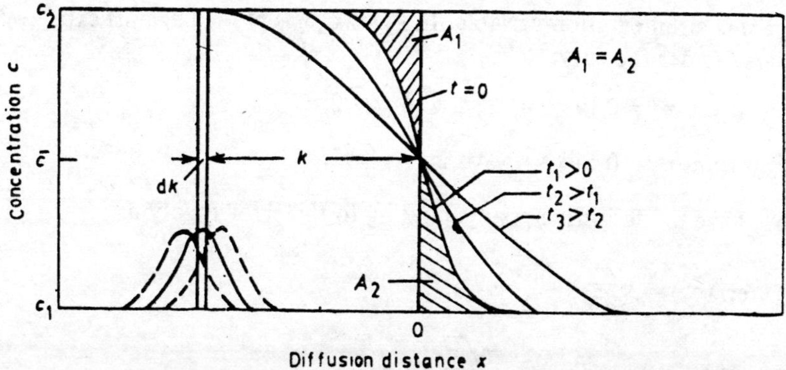

Fig. 2.4 The diffusion couple set-up and the Grube solution.

The cross-over point is the position corresponding to $\bar{c} = (c_1 + c_2)/2$. The origin for the diffusion distance is taken as the cross-over point. This is the only point where the concentration of the diffusing specie decreases on one side (the left side in Fig. 2.4) and increases on the other (the right side).

The Grube solution is simply a summation of a large number of thin film solutions. Consider a thin strip dk at a distance k from the origin. If only this strip were present, the thin film solution would yield:

$$c(x, t) - c_1 = \frac{(c_2 - c_1)\, dk}{2\sqrt{\pi D t}} \exp\left\{-\frac{(x-k)^2}{4Dt}\right\} \tag{2.10}$$

For the entire crystal, the solution is then given by

$$c(x, t) - c_1 = \frac{(c_2 - c_1)}{2\sqrt{\pi D t}} \int_{-\infty}^{0} \exp\left\{-\frac{(x - k^2)}{4Dt}\right\} dk \tag{2.11}$$

Let $(x - k)/2\sqrt{Dt} = \eta$. Then, $-2\sqrt{Dt}\, d\eta = dk$. Using these relationships in Eq. 2.11, we have

$$c(x, t) - c_1 = \frac{c_2 - c_1}{\sqrt{\pi}}\left[-\int_{-\infty}^{0} \exp(-\eta^2)\, d\eta - \int_{0}^{x/2\sqrt{Dt}} \exp(-\eta^2)\, d\eta\right]$$

$$= \frac{c_2 - c_1}{2}\left[1 - \operatorname{erf}\left(\frac{x}{2\sqrt{Dt}}\right)\right] \tag{2.12}$$

The term 'erf' in Eq. 2.12 stands for error function. By definition,

$$\operatorname{erf}\left(\frac{x}{2\sqrt{Dt}}\right) = \frac{2}{\sqrt{\pi}} \int_{0}^{x/2\sqrt{Dt}} \exp(-\eta^2)\, d\eta \tag{2.13}$$

where η is an integration variable. In Fig. 2.5, η is plotted against $\exp(-\eta^2)$. The area under the curve

from $\eta = 0$ to $\eta = +\infty$ is $+\sqrt{\pi}/2$;

from $\eta = 0$ to $\eta = -\infty$ is $-\sqrt{\pi}/2$.

So, we have the following results relating to the error function:

1 $\operatorname{erf}(\infty) = \frac{2}{\sqrt{\pi}} \times \frac{\sqrt{\pi}}{2} = 1$

2 $\operatorname{erf}(-\infty) = \frac{2}{\sqrt{\pi}} \times -\frac{\sqrt{\pi}}{2} = -1$

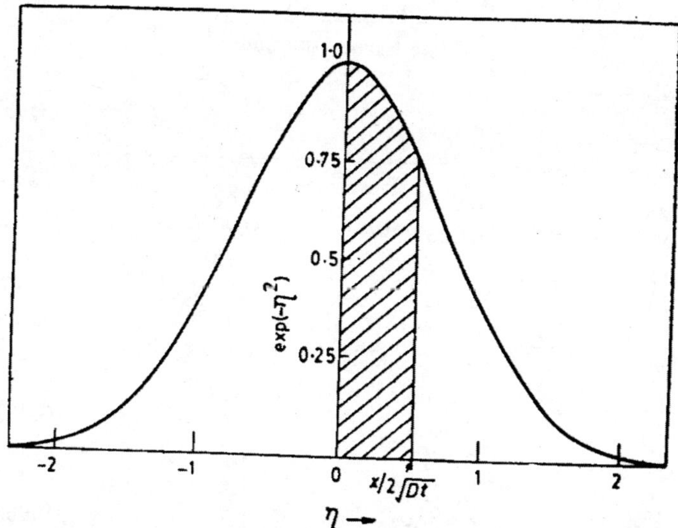

Fig. 2.5 Plot of η versus $\exp(-\eta^2)$. The hatched area is equal to the value of the integral in Eq. 2.13.

3 $\text{erf}\,(0) = 0$

4 $\text{erf}\left(-\frac{x}{2\sqrt{Dt}}\right) = -\text{erf}\left(\frac{x}{2\sqrt{Dt}}\right)$

The initial and boundary conditions of the diffusion couple set-up are:

1 At $x = 0$, for all t, $c = \frac{c_1 + c_2}{2} = \bar{c}$

2 $t = 0$, $x < 0$, $c = c_2$

3 $t = 0$, $x > 0$, $c = c_1$

4 $t > 0$, $x = -\infty$, $c = c_2$

5 $t > 0$, $x = +\infty$, $c = c_1$

6 $t = \infty$, for all x, $c = \frac{+ c_2}{2} = \bar{c}$.

For determining the diffusion coefficient D, the diffusion couple is annealed at a given temperature for a known time t. Then the concentration c is measured as a function of the diffusion distance x with the origin being taken at the cross-over point. The concentration at a given position after the diffusion anneal $c(x, t)$ and the initial concentrations c_1 and c_2 will yield the value of the error function from Eq. 2.12. From the error function Table 2.1, with x and t known, D can be determined.

Even though the Grube solution assumes that D is independent of concentration, it can still be used with reasonable accuracy, when D is a function

TABLE 2.1
The Error Function

z	erf (z)	z	erf (z)
0.000	0.0000	0.85	0.7707
0.025	0.0282	0.90	0.7970
0.05	0.0564	0.95	0.8209
0.10	0.1125	1.0	0.8427
0.15	0.1680	1.1	0.8802
0.20	0.2227	1.2	0.9103
0.25	0.2763	1.3	0.9340
0.30	0.3268	1.4	0.9523
0.35	0.3794	1.5	0.9661
0.40	0.4284	1.6	0.9763
0.45	0.4755	1.7	0.9838
0.50	0.5205	1.8	0.9891
0.55	0.5633	1.9	0.9928
0.60	0.6039	2.0	0.9953
0.65	0.6420	2.2	0.9981
0.70	0.6778	2.4	0.9993
0.75	0.7112	2.6	0.9998
0.80	0.7421	2.8	0.9999

of c, by choosing c_1 and c_2 within a narrow composition range for any one diffusion couple. The experiment can then be repeated with additional diffusion couples to cover the entire composition range.

The Matano-Boltzmann Solution

This solution takes into account the dependence of D on concentration. Therefore, it starts with the more general form of the Fick's second law given by Eq. 2.6:

$$\frac{\partial c}{\partial t} = \frac{\partial}{\partial x}\left(D\frac{\partial c}{\partial x}\right)$$

Boltzmann showed that, if c is a function of a single variable such as $\lambda = x/\sqrt{t}$, Eq. 2.6 can be transformed into an ordinary homogeneous differential equation:

$$-\frac{\lambda}{2}\frac{dc}{d\lambda} = \frac{d}{d\lambda}\left(D\frac{dc}{d\lambda}\right) \tag{2.14}$$

By introducing the initial and boundary conditions for a diffusion couple:

1 $t = 0, \quad x < 0, \quad c = c_2,$

2 $t = 0, \qquad x > 0, \qquad c = c_1$, and

3 $t > 0, \qquad x = \pm\infty, \qquad \dfrac{\partial c}{\partial x} = 0,$

for a fixed annealing time t, the solution of Eq. 2.14 is:

$$D = -\frac{1}{2t}\left(\frac{dx}{dc}\right)\int_{c_1}^{c} x\,dc \tag{2.15}$$

subject to the condition that

$$\int_{c_1} x\,dc = 0 \tag{2.16}$$

This condition is realized by choosing the origin $x = 0$ such that the hatched areas A_1 and A_2 in Fig. 2.6 are equal. The cross-sectional plane corresponding to $x = 0$ is called the *Matano interface*. The cross-over point is not necessarily at $\bar{c}$. The double-hatched area in Fig. 2.6 is the value of the integral in Eq. 2.15.

From experimental data, $c(x)$ is known, so that D can be determined graphically.

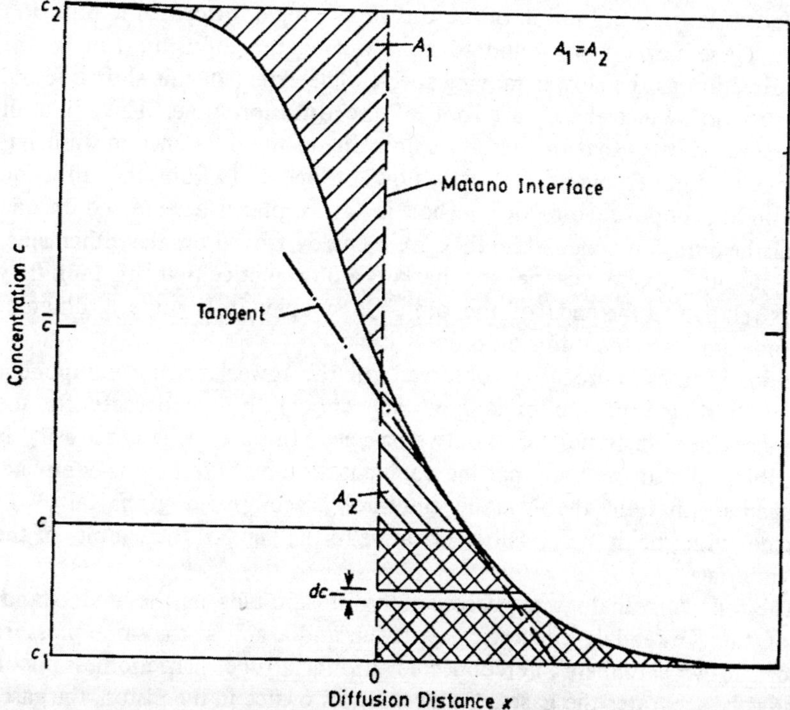

Fig. 2.6 Illustration of the Matano-Boltzmann solution.

2.3 THE INTER-DIFFUSION COEFFICIENT $\tilde{D}$

From Eq. 2.15, it is evident that D depends on the c versus x curve. This curve can result from

1 the B specie diffusing in one direction, or

2 the A specie diffusing in the opposite direction, or

3 both species diffusing simultaneously in directions opposite to each other.

Whether we plot c_A or c_B against x, we end up with the same diffusivity as a function of concentration. This means that D is really a property of the solid solution, without regard to which is the diffusing specie. This diffusivity, usually denoted by $\tilde{D}$, is called the *chemical interdiffusion coefficient*. D_A or D_B is used to refer to the diffusivity of the specie in question and is known as the *intrinsic diffusivity*.

The Kirkendall Effect

In a binary solid solution of A and B, the rates at which A and B diffuse are not necessarily the same. Usually, the lower melting component diffuses faster than the other. This leads to certain interesting effects, as first observed by Smigelskas and Kirkendall.

Inert markers (high melting point, insoluble metals such as Mo and W) are placed at the weld joint of the diffusion couple, prior to the diffusion anneal. These markers are found to shift during the diffusion run in the same direction as the slower-moving specie. The extent of this shift is found to be proportional to the square root of the diffusion time. This kind of movement indicates that the net mass flow due to the difference in diffusivities is being compensated by a bulk flow of matter (within the diffusion zone) in the opposite direction. That is, lattice planes are created on one side of the diffusion zone, while they are being destroyed on the other side. The resulting bulk flow carries the markers along. Notice that this bulk flow occurs relative to the ends of the diffusion couple; it is quite a different phenomenon from the diffusion process itself.

In many cases, porosity is observed on the lower melting component side, indicating that the bulk flow does not fully compensate for the difference in diffusivities of the two species. In cases where porosity is negligible, we can assume that the compensation by bulk flow is complete. For such a situation, the Matano interface as computed graphically will coincide with the initial position, relative to the ends of the couple, of the weld interface.

The following analogy of gaseous interdiffusion aids in the understanding of the Kirkendall effect. Let hydrogen and argon at the same pressure be kept in two chambers interconnected through a tube. A frictionless piston in the tube separates the gases. On opening an orifice in the piston, the gases interdiffuse. The lighter gas, hydrogen, will diffuse faster, resulting in a pressure difference that will tend to shift the piston in the same direction, as the slower diffusing argon is moving.

Darken's Analysis

Darken derived a relationship between $\tilde{D}$, D_A and D_B, using the following assumptions:

1 The molar volume of the solid solutions is independent of concentration,

2 porosity developed during diffusion is negligible, and

3 the bulk flow is perpendicular to the cross-sectional plane, that is, no lateral dimensional changes occur.

Relative to the initial position of the weld interface or the Matano interface, the net flux of atoms is zero. That is,

$$J_A + J_B = 0 \tag{2.17}$$

$$\left(-\tilde{D}\frac{dc_A}{dx}\right) + \left(-\tilde{D}\frac{dc_B}{dx}\right) = 0 \tag{2.18}$$

The fluxes of A and B are due to their intrinsic diffusivities as well as to the bulk flow. Therefore, we can write

$$\left(-D_A\frac{dc_A}{dx} + vc_A\right) + \left(-D_B\frac{dc_B}{dx} + vc_B\right) = 0 \tag{2.19}$$

where v is the velocity of the markers, placed at the original weld. Rearranging Eq. 2.19, we have

$$v = \frac{D_B(dc_B/dx) + D_A(dc_A/dx)}{c_A + c_B} \tag{2.20}$$

As c_A and c_B are concentrations expressed in units of mol m^{-3}, $c_A + c_B = 1/V$, where V is the molar volume assumed to be independent of concentration. So, we have

$$v = (D_B - D_A)\frac{dX_B}{dx} \tag{2.21}$$

where X_B is the mole fraction of B equal to Vc_B and $X_A = 1 - X_B$. Since

$$-\tilde{D}\frac{dc_B}{dx} = -D_B\frac{dc_B}{dx} + vc_B \tag{2.22}$$

we can write

$$-\tilde{D}\frac{dX_B}{dx} = -D_B\frac{dX_B}{dx} + vX_B \tag{2.23}$$

Substituting for dX_B/dx from Eq. 2,23 into 2.21, we get

$$v = (D_B - D_A)\left(\frac{vX_B}{D_B - \tilde{D}}\right) \tag{2.24}$$

or

$$D_B - \tilde{D} = (D_B - D_A)X_B \tag{2.25}$$

$$\tilde{D} = D_A X_B + D_B X_A \tag{2.26}$$

which is the relation between the interdiffusion coefficient and the intrinsic diffusivities derived by Darken. It can be determined from the experimental shift of the markers, noting that the shift is proportional to the square root of time. $\tilde{D}$ can be determined from the Matano-Boltzmann analysis. Using Eqs. 2.21 and 2.26, D_A and D_B can be calculated.

2.4 ATOMIC THEORY OF DIFFUSION

Fick's First Law from the Atomic Model

In a crystal, consider two adjacent interatomic planes separated by a distance δ and perpendicular to the diffusion direction x. Let there be n_1 moles of the diffusing specie per unit area in plane 1 and n_2 in plane 2, with $n_1 > n_2$. If ν' is the frequency with which atoms jump from one plane to a neighbouring plane (the jump may be in either the forward direction or the backward direction),

$$J_{1\to 2} = \tfrac{1}{2} n_1 \nu' \tag{2.27}$$

$$J_{2\to 1} = \tfrac{1}{2} n_2 \nu' \tag{2.28}$$

$$J_{1\to 2(\text{net})} = \tfrac{1}{2} \nu' (n_1 - n_2) \tag{2.29}$$

If $n_1 = c_1\delta$ and $n_2 = c_2\delta$, and $(c_1 - c_2) = -\delta \mathrm{d}c/\mathrm{d}x$, we have

$$J_{\text{net}} = -\tfrac{1}{2}\delta^2 \nu' \frac{\mathrm{d}c}{\mathrm{d}x} \tag{2.30}$$

Comparing Eq. 2.30 with Fick's first law, we see that

$$D = \tfrac{1}{2}\delta^2 \nu' \tag{2.31}$$

If we take into account the probability of jumps in three mutually perpendicular directions, we can rewrite

$$D = \frac{1}{6}\delta^2 \nu' \tag{2.32}$$

Mechanisms of Diffusion

Atoms which are small enough to occupy interstitial sites diffuse by jumping from one interstitial site to a neighbouring interstitial site. This is known as the *interstitial mechanism* of diffusion, Fig. 2.7(a). In dilute interstitial phases, the probability that an interstitial atom will find a vacant interstitial site in the diffusing direction is high and can be taken to be unity.

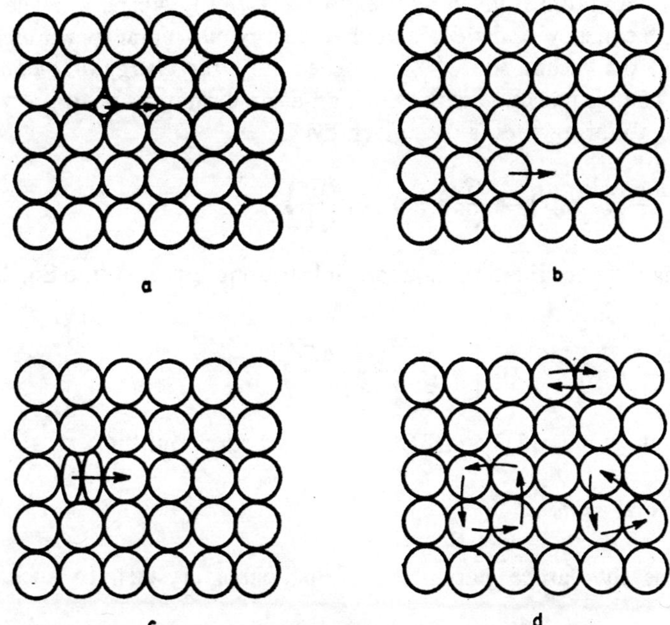

Fig. 2.7 (a) Interstitial, (b) vacancy, (c) interstitialcy, (d) direct exchange and ring mechanisms of diffusion.

Substitutional atoms can diffuse by interchanging positions with a neighbouring vacant site, giving rise to the *vacancy mechanism* of diffusion, Fig. 2.7(b). The probability of a diffusing atom finding such a vacancy is proportional to the vacancy concentration in the crystal. Vacancy concentrations in thermal equilibrium are of the order of 10^{-4} near the melting point of the crystal and decrease by several orders of magnitude at lower temperatures.

In the *interstitialcy mechanism*, Fig. 2.7(c), an atom from a regular site moves into an interstitial site and displaces another atom from a regular site to an interstitial site. In effect, the two atoms share a common lattice site, but are both displaced from it.

In the *direct exchange* of two neighbouring atoms, Fig. 2.7(d), severe local distortion is involved. Thus this mechanism is not considered likely.

In the *ring mechanism*, three or four atoms in the form of a ring rotate, thereby exchanging their positions, Fig. 2.7(d). This process, when repeated again and again with a different set of atoms in the ring each time, can move an atom far away from its initial position.

Temperature Dependence of D

For a substitutional atom diffusing via the vacancy mechanism, we have to consider the availability of a neighbouring vacant site for the atom to jump into. The probability that the diffusing atom will find a particular neighbouring site to be vacant is equal to $\exp(-\Delta G_f/RT)$, where ΔG_f is the free energy of formation of a mole of vacancies. The number of successful jump attempts by an atom is given by $\nu \exp(-\Delta G_m/RT)$, where ν is the lattice vibration frequency and ΔG_m is the free energy maximum (per mole) along the path to the vacant site. ΔG_m is called the *free energy of motion* of a vacancy. The frequency ν' with which an atom exchanges position with any of the neighbouring sites is then given by

$$\nu' = Z\nu \exp\left(-\frac{\Delta G_m}{RT}\right) \exp\left(-\frac{\Delta G_f}{RT}\right) \tag{2.33}$$

where Z is the coordination number. Substituting Eq. 2.33 into Eq. 2.32, we obtain

$$D = \frac{1}{6}\delta^2 Z\nu \exp\left[-\left(\frac{\Delta G_m + \Delta G_f}{RT}\right)\right] \tag{2.34}$$

In simple cubic, FCC and BCC crystals, it turns out that

$$\frac{1}{6}\delta^2 Z = a_0^2 \tag{2.35}$$

where a_0 is the lattice parameter of the cubic crystals. In such cases, therefore,

$$D = a_0^2\nu \exp\left[-\left(\frac{\Delta G_m + \Delta G_f}{RT}\right)\right] \tag{2.36}$$

Using $\Delta G = \Delta H - T\Delta S$, we can write Eq. 2.36 as

$$D = a_0^2\nu \exp\left(\frac{\Delta S_m + \Delta S_f}{R}\right) \exp\left[-\left(\frac{\Delta H_m + \Delta H_f}{RT}\right)\right] \tag{2.37}$$

Experimental data show that

$$D = D_0 \exp\left(-\frac{Q}{RT}\right) \tag{2.38}$$

Values of D_0 (called the *frequency factor*) and Q (called the *activation energy for diffusion*) are obtained from measurements of D at different temperatures, for example, by using a diffusion couple set-up. A plot of $\ln D$ versus $1/T$ yields a straight line. The slope of the line is equal to $-Q/R$ and the intercept of the y-axis is $\ln D_0$.

Comparing Eqs. 2.37 and 2.38, it is seen that

$$Q = \Delta H_m + \Delta H_f \tag{2.39}$$

and

$$D_0 = a_0^2\nu \exp\left(\frac{\Delta S_m + \Delta S_f}{R}\right) \tag{2.40}$$

Approximate values of D_0 and Q for some diffusion processes are listed in Table 2.2.

TABLE 2.2
Approximate D_0 and Q Values for Some Diffusion Processes

Diffusion process	D_0 10^{-4} m^2 s^{-1}	Q kJ mol^{-1}
Cu in Cu	0.20	196
Zn in Zn	0.15	94
Al in Al	1.98	143
Fe in Fe (α)	118	281
Ge in Ge	9.3	288
Si in Si	5400	477
W in W	43	640
C in graphite	7	681
H in Fe	0.001	13
N in Fe	0.005	75
C in Fe (α)	2.2	122
C in Fe (γ)	0.2	142
V in Fe	3.9	244
Mn in Fe	4.0	305
Ni in Fe	2.6	295
Zn in Cu	0.73	170
Ni in Cu	2.0	230
Cu in Al	0.25	121

For interstitial diffusion in a dilute interstitial solution, the jump frequency into a given neighbouring interstitial site is $\nu \exp(-\Delta G_m/RT)$. Here, ΔG_m represents the free energy increase, as the diffusing atom moves from one interstitial site to the next. It is called the free energy of motion of an interstitial. The probability that the adjacent site will be vacant is almost unity. Then the following expression for D as a function of temperature results:

$$D = \Gamma a_0^2 \nu \exp\left(-\frac{\Delta G_m}{RT}\right) \tag{2.41}$$

where Γ is a constant that depends on the number and geometry of the interstitial sites in the cubic unit cell.

Huntington and Seitz have estimated from theoretical considerations the values of ΔH_m for various diffusion mechanisms in copper. These are listed in Table 2.3.

These values indicate that the most probable mechanism of diffusion in copper is the vacancy mechanism. It should also be emphasized that the direct interchange and the ring mechanisms of diffusion cannot explain the Kirkendall effect observed in solid solutions, where the two different species diffuse at different rates.

TABLE 2.3

ΔH_m for Different Diffusion Mechanisms in Copper

Mechanism	ΔH_m kJ mol^{-1}
Direct interchange	840
Interstitialcy	880
Vacancy	120
4-ring mechanism	370

Table 2.4 compares the values of ΔH_f and ΔH_m with the experimental values of Q for self diffusion in silver and gold. It is seen that $\Delta H_f + \Delta H_m \approx Q$.

TABLE 2.4

Calculated and Experimental Activation Energies for Diffusion by the Vacancy Mechanism (values in kJ mol^{-1})

Element	ΔH_f	ΔH_m	$\Delta H_f + \Delta H_m$	Q
Silver	97	80	177	174
Gold	95	79	174	184

There is no probability factor involving ΔH_f for interstitial diffusion, see Eq. 2.41. This fact is responsible for interstitial diffusion generally being much faster than substitutional diffusion occurring by the vacancy mechanism. For example, the diffusion coefficient of carbon in FCC iron at 1000°C is 3×10^{-11} m^2 s^{-1}, while that of nickel in FCC iron at the same temperature is much less, 2×10^{-16} m^2 s^{-1}.

2.5 OTHER DIFFUSION PROCESSES

Diffusion in Ionic Crystals

In ionic crystals, point defects in thermal equilibrium occur in pairs, so that the electrical charge neutrality of the crystal as a whole is maintained. A pair of one cation vacancy and one anion vacancy is known as a *Schottky defect*. A pair of a cation vacancy and a cation interstitial is called a *Frenkel defect*. When Schottky defects dominate in an ionic crystal, the cation vacancy usually carries the diffusional flux. When Frenkel defects are abundant, the cation interstitial carries the flux.

In addition to these defects in thermal equilibrium, ionic crystals may have defects generated by impurities. If a Cd^{2+} cation replaces two Na^+

cations in a NaCl crystal, such that the overall electrical neutrality is maintained, a cation vacancy is created. Therefore, the addition of Cd^{2+} will increase the diffusivity, since Na^+ vacancies carry the flux here. A fraction of a percent of $CdCl_2$ is sufficient to produce cation vacancies far in excess of the number in thermal equilibrium.

In the ionic crystal, defects can also be generated by deviations from stoichiometry, resulting in an increase in the diffusional flux. In zinc oxide, the Frenkel defects dominate and the flux is carried by cation interstitials. Off-stoichiometric compounds such as Zn_xO, where $x > 1$ have a larger concentration of cation interstitials than a stoichiometric compound, Fig. 2.8(a). In iron oxide (FeO), the Schottky defects dominate and the flux is carried by cation vacancies. Off-stoichiometric compounds, such as Fe_yO, where $y < 1$, have a larger number of cation vacancies than the stoichiometric compound, Fig. 2.8(b).

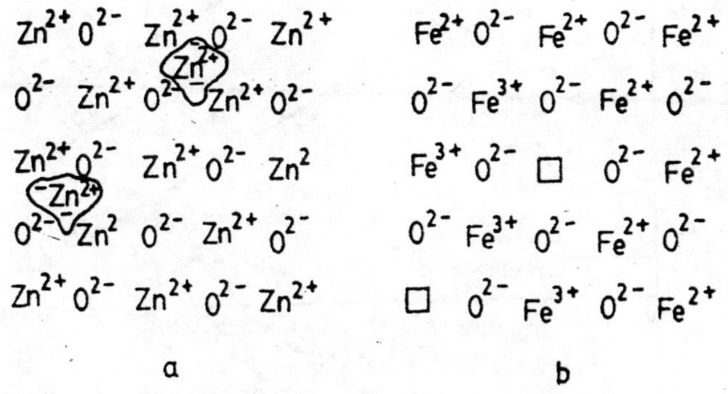

Zn^{2+} O^{2-} Zn^{2+} O^{2-} Zn^{2+}
(Zn^{+})
O^{2-} Zn^{2+} O^{2-} Zn^{2+} O^{2-}
Zn^{2+} O^{2-} Zn^{2+} O^{2-} Zn^{2}
(Zn^{2+})
O^{2-} Zn^{2} O^{2-} Zn^{2+} O^{2-}
Zn^{2+} O^{2-} Zn^{2+} O^{2-} Zn^{2+}

a

Fe^{2+} O^{2-} Fe^{2+} O^{2-} Fe^{2+}
O^{2-} Fe^{3+} O^{2-} Fe^{2+} O^{2-}
Fe^{3+} O^{2-} □ O^{2-} Fe^{2+}
O^{2-} Fe^{3+} O^{2-} Fe^{2+} O^{2-}
□ O^{2-} Fe^{3+} O^{2-} Fe^{2+}

b

Fig. 2.8 Defects present in (a) ZnO and (b) FeO, owing to deviations from stoichiometry. Vacant cation sites in FeO are indicated by □.

Measurements of the electrical conductivity of ionic crystals yield diffusivity from the Nernst-Einstein relation:

$$\frac{\sigma}{D} = \frac{ne^2z}{kT} \tag{2.42}$$

where

σ is the electrical conductivity due to the diffusing defect,
n is the number of diffusing defects per unit volume,
z is their valence, and
e is the electronic charge.

Figure 2.9 is a plot of conductivity versus $1/T$ for a NaCl crystal doped with varying atomic percents of $CdCl_2$. At temperature above 650°C, the thermally-generated cation (Na^+) vacancies dominate and the conductivities of the pure and doped NaCl crystals are the same. At lower temperatures,

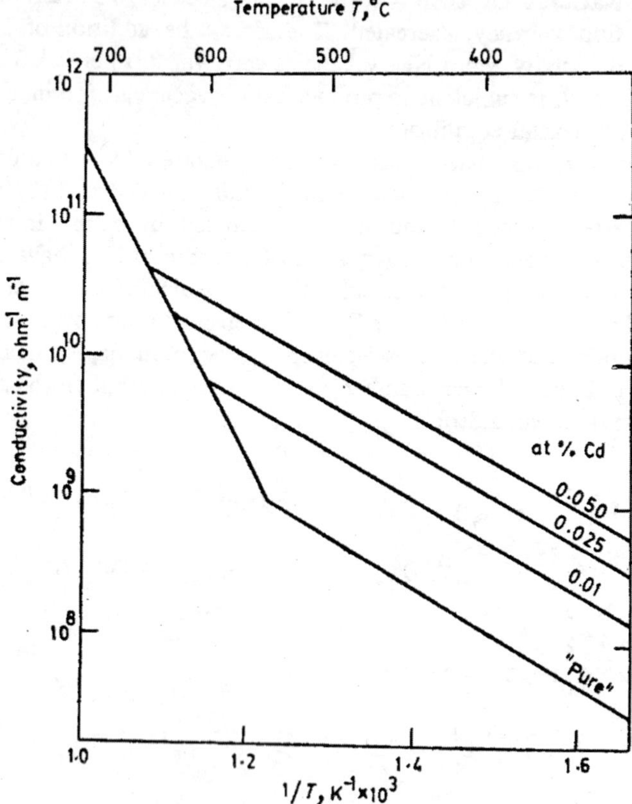

Fig. 2.9 Electrical conductivity of NaCl doped with different amounts of $CdCl_2$.

the conductivity is a function of the degree of doping. At any one temperature in this range, the pure crystals have the lowest conductivity. With increasing $CdCl_2$ concentration, the impurity-generated vacancies increase in number, with a corresponding increase in conductivity.

Diffusion along Grain Boundaries

In a polycrystalline material, the grain boundary regions are not as closely packed as the crystal. The activation enthalpy of motion ΔH_m for grain boundary diffusion is correspondingly smaller than that for lattice diffusion. The experimental activation energies are likewise smaller, i.e., $Q_{gb} < Q_{lattice}$.

A lower activation energy does not necessarily mean that diffusion along high diffusivity paths such as grain boundaries will always dominate the lattice diffusion. The cross-sectional area across which mass transport can occur (see Eq. 2.1) is usually much smaller for special diffusion paths than for lattice diffusion.

The diffusion coefficients for self-diffusion determined for a single crystal and for polycrystalline silver are shown in Fig. 2.10 as a function of $1/T$. At

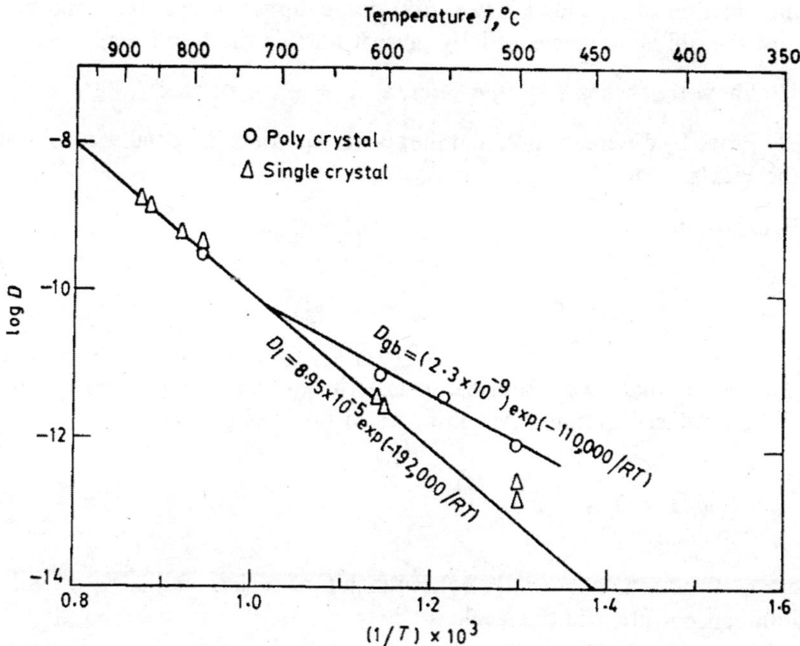

Fig. 2.10 Diffusion coefficient D for self-diffusion in a single crystal and polycrystalline silver. The units of D_0 and Q are $m^2\ s^{-1}$ and $kJ\ mol^{-1}$.

lower temperatures, diffusion in the polycrystalline material is faster than in the single crystal, as grain boundary diffusion is dominant here. At higher temperatures, lattice diffusion is dominant, so that the diffusion coefficient is the same here for both single crystal and polycrystalline silver.

FURTHER READING

P.G. Shewmon, *Diffusion in Solids*, McGraw-Hill, New York (1963).

C. Matano. *Japan, J. Phys.*, **8**, 109 (1933).

A. Smigelskas and E. Kirkendall, *Trans. Amer. Inst. Min. Met. Engrs.*, **171**, 130 (1947).

L. Darken, *Trans. Amer. Inst. Min. Met. Engrs.*, **174**, 184 (1948).

H.B. Huntington and F. Seitz, *Phys. Rev.*, **61**, 315 (1942).

EXERCISES

2.1 The solution of Fick's second law for the thin-film case is

$$c - c_l = \frac{\alpha}{2\sqrt{\pi D t}} \exp\left(-\frac{x^2}{4Dt}\right)$$

where D is not a function of the concentration, α is the number of moles of the diffusing specie per m^2 initially present in the form of a very thin layer at the junction of a sandwich type diffusion couple and c_1 is the concentration of the diffusing specie initially present outside the thin layer.

1 Show that when $t > 0$, $c \to c_1$, as $x \to +\infty$ or as $x \to -\infty$.

2 Prove by differentiation that the above equation is actually a solution of the Fick's second law.

3 Prove that

$$\alpha = \int_{-\infty}^{+\infty} (c - c_1)\, dx$$

2.2 Show that, in a carburization experiment, the total quantity of moles of carbon diffused in time t per unit area of the steel surface is

$$\int_0^t J\, dt = 2(c_s - c_1)\sqrt{\frac{Dt}{\pi}}$$

where c_s is the constant surface concentration of carbon and c_1 is the initial carbon concentration in the steel.

2.3 A rod of pure copper was joined to a rod of a Cu-29.4 at.% Zn alloy. After annealing for 360 hr, the % Zn versus distance data came out to be as follows:

at.% Zn	*x, mm*	*at.% Zn*	*x, mm*
0.3	5.005	23.5	3.655
1.5	4.815	25.0	3.405
4.4	4.645	26.5	3.075
8.8	4.495	27.9	2.515
14.7	4.315	28.8	1.895
20.6	3.965	29.1	1.495

Distances were measured from a fixed point outside the diffusion zone. Determine the position of the Matano interface and calculate $D(c)$ at 5, 15 and 25 at.% Zn.

2.4 In a Kirkendall diffusion experiment analyzed by the Matano method, the following data are obtained for the metal at the cross-section containing the markers:

Diffusion time	$t = 200$ hrs
Marker movement	$s = 1.44$ mm
Interdiffusion coefficient	$\tilde{D} = 10^{-11}$ m^2 s^{-1}

Slope of the penetration curve at the markers $\frac{\partial c_B}{\partial x} = 0.2\ \text{mm}^{-1}$

Atom fraction of A $= 0.4$

Compute D_A and D_B.

2.5 In copper at 1000°C, calculate the equilibrium fractions of (i) interstitial atoms, and (ii) vacancies, using Huntington's values for ΔH_f (interstitial) $=$ 880 kJ mol^{-1} and ΔH_f (vacancies) $=$ 120 kJ mol^{-1}.

2.6 An Al-4% Cu alloy is heated to 550°C during heat treatment and quenched to room temperature. Immediately after the quench, the diffusion rate of copper (which proceeds by a vacancy mechanism) was found to be 10^7 times faster than what would be expected from the listed diffusion data. What fraction of vacancies in equilibrium at 550°C is retained at room temperature by the rapid quenching? The enthalpy of motion of vacancy in this alloy is 50 kJ mol^{-1}.

2.7 Find the grain size of a polycrystalline solid for the same amount of material to be transported through (i) the grain and (ii) the grain boundary at 500°C. Assume that the grains are cube shaped and the grain boundaries are 5Å thick.

For lattice diffusion: $D_0 = 0.7 \times 10^{-4}\ \text{m}^2\ \text{s}^{-1}$

$Q = 188\ \text{kJ mol}^{-1}$

For grain boundary diffusion: $D_0 = 0.09 \times 10^{-4}\ \text{m}^2\ \text{s}^{-1}$

$Q = 90\ \text{kJ mol}^{-1}$

2.8 What do you understand by intrinsic and extrinsic regions of diffusion in a sodium chloride crystal doped with a divalent cation impurity? Explain how the enthalpies of formation and motion of Na^+ ions can be evaluated from the experimental determination of electrical conductivity versus temperature.

3

Thermodynamics of Transformations

A phase transformation can occur spontaneously only when the free energy change during the transformation is negative. In order to know whether this condition is satisfied for a transformation, we need to know the free energy of the parent and the product phases. The free energies of elemental crystals and solid solutions are discussed here as a function of composition and the external parameters: temperature and pressure. The thermodynamic order of transformations is defined. It is shown how a driving force arises for first-order transformations with and without compositional changes. Finally, second-order transformations are briefly discussed.

3.1 FREE ENERGY OF ELEMENTAL CRYSTALS

In elements, the composition is not a variable. In addition, we shall assume that the elemental crystals are perfect, so that there is no configurational entropy associated with crystal imperfections. Then, the variables that are to be considered are the external variables, e.g., temperature and pressure.

The Gibbs free energy G of the crystal is given by

$$G = H - TS \tag{3.1}$$

where H and S are the enthalpy and the entropy of the crystal and T is the absolute temperature (in kelvin). The enthalpy term can be written as

$$H = H_0 + \int_0^T c_p \, \mathrm{d}T \tag{3.2}$$

where H_0 is the residual enthalpy at 0 K and c_p is the heat capacity at constant pressure. The entropy of the crystal is given by

$$S = \int_0^T \frac{c_p \, \mathrm{d}T}{T} \tag{3.3}$$

Combining Eqs. 3.1 through 3.3, we obtain at constant pressure

$$\left(\frac{\partial G}{\partial T}\right)_P = -S \tag{3.4}$$

Integrating Eq. 3.4 within appropriate limits, we can write

$$\int_{H_0}^{G} dG = \int_0^T S\, dT \tag{3.5}$$

$$G = H_0 - \int_0^T S\, dT$$

$$= H_0 - \int_0^T \left(\int_0^T \frac{c_p\, dT}{T} \right) dT \tag{3.6}$$

Figure 3.1(a) is a schematic plot of the variation in the Gibbs free energy as a function of temperature at constant pressure. The curve has zero slope at 0 K, the value at that temperature being the residual enthalpy or bond energy H_0. The slope becomes more and more negative, as the temperature is increased. The rate at which the slope decreases with temperature is dependent on the value of the heat capacity, which is usually a function of temperature.

From the thermodynamic laws, we can show that

$$\left(\frac{\partial G}{\partial P}\right)_T = V \tag{3.7}$$

Thus, the Gibbs free energy variation with pressure at a constant temperature has a positive slope, equal to the molar volume V, as illustrated schematically in Fig. 3.1(b).

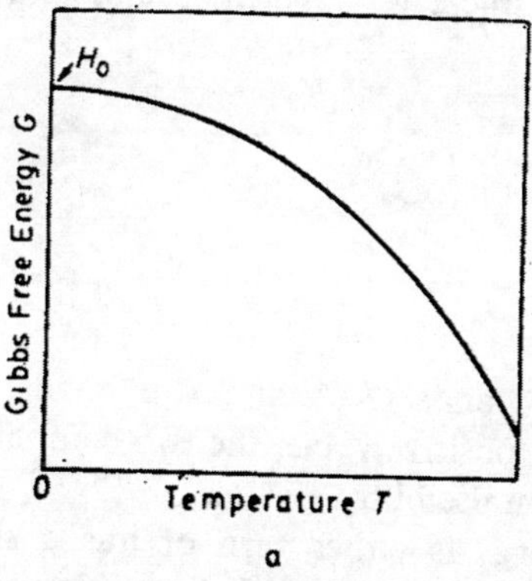

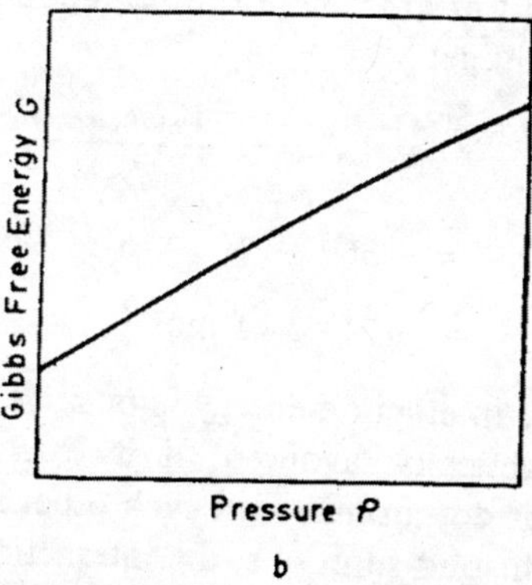

Fig. 3.1 Schematic variation of the Gibbs free energy G as a function of (a) temperature at constant pressure, and (b) pressure at constant temperature.

The above description of the free energy of elemental crystals also applies to chemical compounds with ideal stoichiometry, for which the configurational entropy is zero.

3.2 FREE ENERGY OF SOLID SOLUTIONS

Here, the composition is a variable. So, the contribution to the free energy from the configurational entropy cannot be ignored.

The Interaction Energy

The bond energy of a solid solution depends on the type of bonds between nearest-neighbour atoms in the crystal. We shall ignore the energy contributions to the bond energy from next-nearest neighbours. The types of bonds present in a binary solid solution of A and B atoms are A-A, B-B and A-B bonds. Let V_{AA}, V_{BB} and V_{AB} be the energies of these bonds respectively. When pure A is dissolved in pure B, some of the A-A and B-B bonds are broken to create new A-B bonds:

$$\begin{matrix} A\text{—}A \\ \\ B\text{—}B \end{matrix} \quad \rightarrow \quad \begin{matrix} A & A \\ | & | \\ B & B \end{matrix}$$

The interaction energy $\bar{V}$ is defined as the energy change that occurs when A-A and B-B bonds are broken to produce one A-B bond. As illustrated above, when one A-A and one B-B bond are broken, two A-B bonds are produced. So, we have

$$\bar{V} = V_{AB} - \left(\frac{V_{AA} + V_{BB}}{2}\right) \tag{3.8}$$

If the interaction energy is negative, unlike bonds (A-B bonds) are energetically favoured. If $\bar{V}$ has a large negative value, compound formation is favoured, where every A atom is surrounded by B neighbours and vice versa. For example, in the compound of silica, $\bar{V}$ is negative and large in magnitude:

$$\bar{V} = V_{Si\text{-}O} - \frac{V_{Si\text{-}Si} + V_{O\text{-}O}}{2}$$

$$= -369 - \tfrac{1}{2}(-176 - 139)$$

$$= -211.5 \text{ kJ mol}^{-1}.$$

If the interaction energy is positive, like bonds (A-A and B-B bonds) are energetically favoured. If it is a large positive value, the two component atoms do not mix with each other and immiscibility results.

In solid solutions, the interaction energy is either zero or has a small positive or negative value. If $\bar{V}$ is zero, we have a perfectly random *ideal solution*. If $\bar{V}$ is negative, the tendency will be for short range ordering in the solid solution. *Short range ordering* refers to formation of small regions in the solution, where A-B bonds are present preferentially. If $\bar{V}$ is positive, the tendency will be for clustering of like atoms in the solid solution. *Clustering* refers to the formation of small regions in the solid solution, where A-A or B-B bonds are preferentially present.

Clustering or Short-range Order Parameter α

Let $P_{A(B)}$ be the probability that a B atom will have an A atom as the nearest neighbour on a specified atomic site. In a truly random solid solution, there is no preference for any type of bonds. So, we have here

$$P_{A(B)} = X_A \tag{3.9}$$

where X_A is the mole fraction of A in the solution. Where the tendency is for short-range ordering, the B atom will prefer to have A neighbours. So, for this solution,

$$P_{A(B)} > X_A \tag{3.10}$$

Where there is a tendency for clustering, the B atom will prefer B neighbours, as B-B bonds lower the energy. So, here

$$P_{A(B)} < X_A \tag{3.11}$$

The local ordering (or clustering) parameter α is defined as

$$\alpha = 1 - \frac{P_{A(B)}}{X_A} \tag{3.12}$$

1 If $P_{A(B)} = X_A$, $\alpha = 0$, we have a truly random solution.
2 If $P_{A(B)} > X_A$, α is negative, the tendency is for ordering.
3 If $P_{A(B)} < X_A$, α is positive, the tendency is for clustering.

The number of A-B bonds, N_{AB}, in one mole of the solid solution is given by

$$N_{AB} = P_{A(B)} Z N_0 X_B \tag{3.13}$$

where

Z is the coordination number (number of nearest-neighbours of an atom),

N_0 is Avogadro's number, and

X_B is the mole fraction of B.

Here, $P_{A(B)}Z$ gives the number of A atoms on the Z sites surrounding a B atom. N_0X_B is the number of B atoms in one mole of the solid solution. Substituting for $P_{A(B)}$ from Eq. 3.12, we obtain

$$N_{AB} = Z N_0 X_A X_B (1 - \alpha) \tag{3.14}$$

Free Energy as a Function of Temperature

At 0 K, the bond energy H_0 will be the sum of the energies of the three types of bonds in the solid solution:

$$H_0 = N_{AA}V_{AA} + N_{BB}V_{BB} + N_{AB}V_{AB} \tag{3.15}$$

where N_{AA}, N_{BB} and N_{AB} are the number of A-A, B-B and A-B bonds. In order to evaluate H_0, we can consider the bond energies of pure A and pure B before mixing and the energy change that occurs on mixing:

$$H_0 = \underbrace{\tfrac{1}{2}ZN_0X_AV_{AA} + \tfrac{1}{2}ZN_0X_BV_{BB}}_{\text{energy before mixing}} + \underbrace{ZN_0X_AX_B(1-\alpha)\bar{V}}_{\text{energy of mixing}}$$

$$= \tfrac{1}{2}ZN_0[X_AV_{AA} + X_BV_{BB} + 2X_AX_B(1-\alpha)\bar{V}] \tag{3.16}$$

The *configurational entropy* S_c increases on mixing. It can be evaluated on the assumption that the solid solution is truly random:

$$S_c = -R(X_A \ln X_A + X_B \ln X_B) \tag{3.17}$$

This assumption is an approximation in cases where there is some tendency for short range ordering or clustering in the solid solution. The treatment based on this assumption is known as the *regular solution model*.

The Gibbs free energy G of the solid solution at temperature T is then given by:

$$G = H - T(S_{\text{thermal}} + S_{\text{configurational}})$$

$$= \tfrac{1}{2}ZN_0[X_AV_{AA} + X_BV_{BB} + 2X_AX_B(1-\alpha)\bar{V}]$$

$$+ \int_0^T c_p^s \, dT - T\int_0^T \frac{c_p^s \, dT}{T} + RT(X_A \ln X_A + X_B \ln X_B) \tag{3.18}$$

where c_p^s is the heat capacity of the solid solution. We can assume that the heat capacity of the solid solution varies according to the *Neumann-Kopp's rule*:

$$c_p^s = c_p^A X_A + c_p^B X_B \tag{3.19}$$

A straight line relationship results, when c_p^s is plotted as a function of composition.

The entropy of mixing S_c and $-TS_c$ at constant temperature vary as shown in Fig. 3.2. The $-TS_c$ curve has a U-shape. As the composition approaches that of one of the pure components, the magnitude of the slope of the $-TS_c$ curve tends to infinity.

The general variation of free energy as a function of composition at different temperatures can be evaluated for each of the following three cases.

Ideal Solid Solution: $\bar{V} = 0$, $\alpha = 0$

In Eq. 3.18, the term containing the interaction energy $\bar{V}$ is zero for this case. Also, at 0 K, the terms containing the configurational entropy and the heat capacity are zero. So, the Gibbs free energy G, as given by Eq. 3.18,

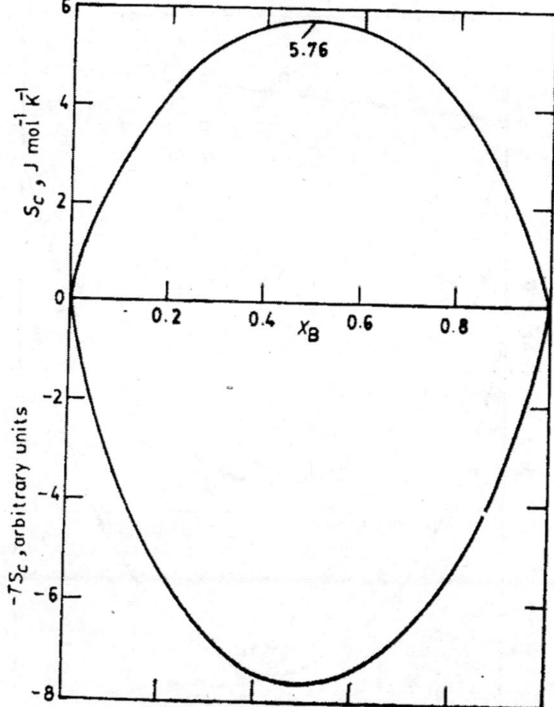

Fig. 3.2 (a) Configurational entropy S_c, and (b) $-TS_c$, as a function of composition in a binary random solid solution.

varies linearly with composition at 0 K, Fig. 3.3. At $T > 0$ K, the configurational entropy term $-TS_c$ has a U-shape as a function of composition as shown in Fig. 3.2. When this is added to the bond energy terms and the heat capacity terms in Eq. 3.18, a U-curve still results as shown in Fig. 3.3. As the temperature increases, the curve becomes more and more U-shaped, with increasing dominance of $-TS_c$, Fig. 3.3.

Tendency for Ordering: $\bar{V} < 0$, $\alpha < 0$

At $T = 0$ K, by virtue of the term containing the negative interaction energy $\bar{V}$ in Eq. 3.18, the free-energy composition curve has a shallow U-shape, Fig. 3.4. This becomes more and more U-shaped, as the temperature increases and the configurational entropy term $-TS_c$ becomes increasingly dominant, Fig. 3.4.

For such a binary system, at very high temperatures, disorder prevails so that the configurational entropy term is maximized. On lowering the temperature, the tendency for short-range ordering increases. Eventually, *long-range order* sets in and an ordered phase becomes stable.

Tendency for Clustering and Immiscibility: $\bar{V} > 0$, $\alpha > 0$

As the interaction energy $\bar{V}$ in Eq. 3.18 is positive in this case, the free-

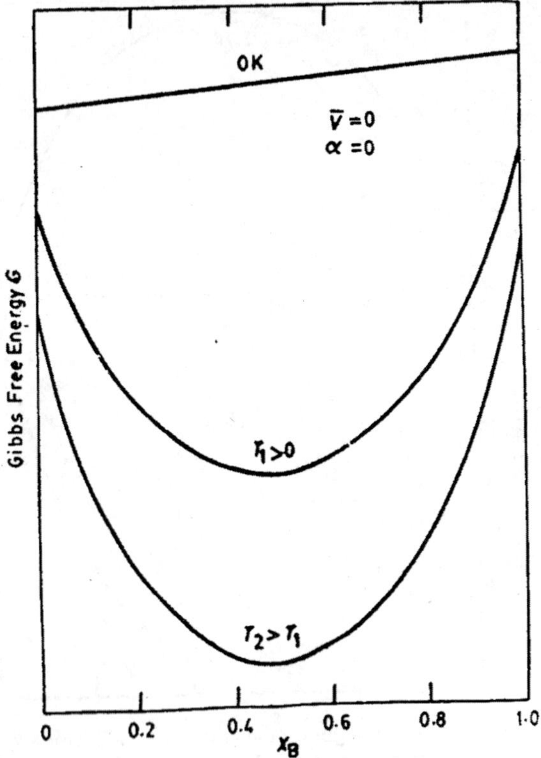

Fig. 3.3 Variation of Gibbs free energy G with composition at different temperatures in an ideal binary solid solution.

energy composition curve has a shallow inverse-U (∩) shape at 0 K, Fig. 3.5. At $T > 0$ K, the configurational entropy term has a U-shape; the addition of these two contributions to the free energy results in a curve with two minima and one maximum, Fig. 3.5. As the temperature increases, the $-TS_c$ term in Eq. 3.18 dominates and the free-energy composition curve takes on a U-shape, see Fig. 3.5.

For such a binary system, at high temperatures, complete disorder prevails and the tendency for clustering is negligible. However, this tendency increases, as the temperature is lowered. Eventually, *the miscibility gap* appears on the phase diagram. Separation of two phases, one poor in B component and the other rich in B component, occurs.

3.3 THERMODYNAMIC ORDER OF TRANSFORMATIONS

Transformations can be classified as first order, second order, etc. depending on how the Gibbs free energy G changes with the external parameters of temperature and pressure.

First Order Transformations

Here, the Gibbs free energy changes continuously through a phase trans-

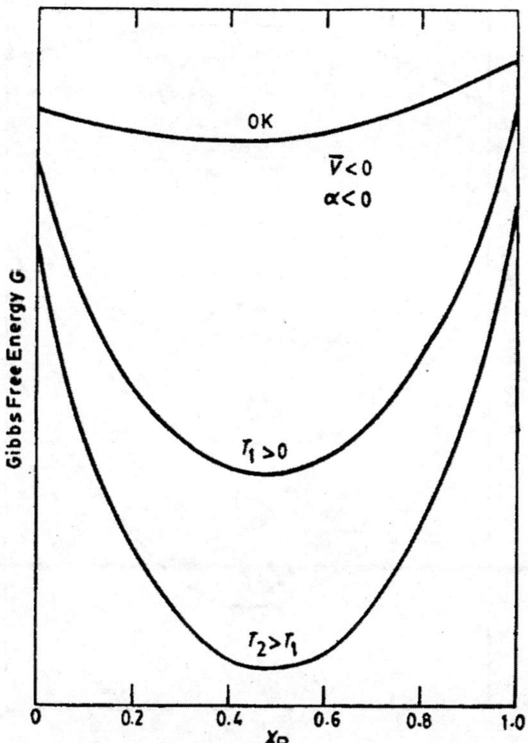

Fig. 3.4 Variation of Gibbs free energy G with composition at different temperatures in a binary solid solution with tendency for ordering.

formation at the equilibrium temperature T_0 and pressure P_0 as illustrated in Fig. 3.6 for an $\alpha \rightarrow \beta$ transformation. However, the free energy curve changes slope abruptly at T_0 or P_0. That is, the first derivatives of G with respect to temperature and pressure have discontinuities. From Eqs. 3.4 and 3.7, we note that

$$\left(\frac{\partial G}{\partial T}\right)_P = -S$$

and

$$\left(\frac{\partial G}{\partial P}\right)_T = V$$

So, first order transformations are associated with a volume change and an entropy change.

Second Order Transformations

A second order transformation is one in which the free energy G and its first derivatives change continuously through the equilibrium transition

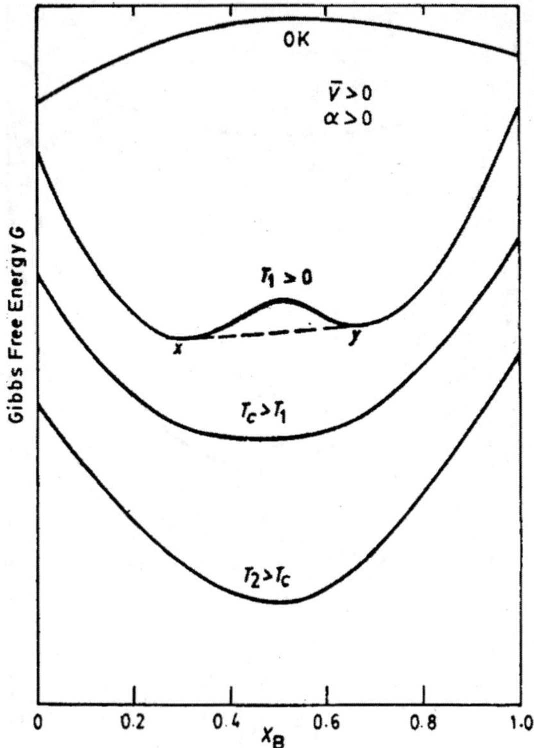

Fig. 3.5 Variation of Gibbs free energy G with composition at different temperatures in a binary solid solution with tendency for clustering and immiscibility.

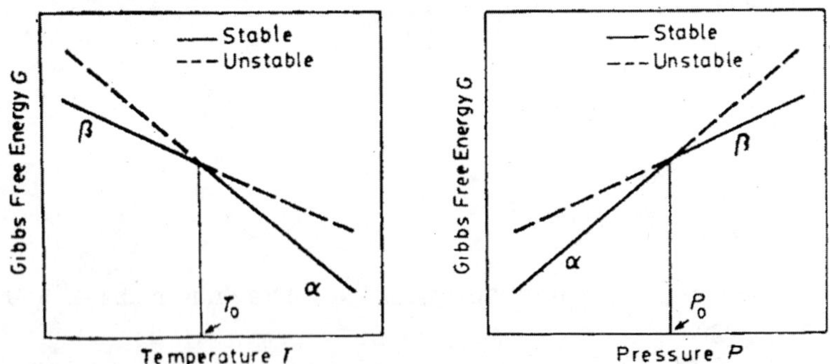

Fig. 3.6 Gibbs free energy G of the α and β phases as a function of (a) temperature at constant pressure, and (b) pressure at constant temperature.

temperature T_c and pressure P_c, but the second derivatives have discontinuities. Through the relationships

$$\left(\frac{\partial^2 G}{\partial T^2}\right) = -\left(\frac{\partial S}{\partial T}\right)_P \tag{3.20}$$

$$\left(\frac{\partial^2 G}{\partial P^2}\right) = \left(\frac{\partial V}{\partial P}\right)_T \tag{3.21}$$

and

$$\left(\frac{\partial^2 G}{\partial T\,\partial P}\right) = \left(\frac{\partial V}{\partial T}\right)_P \tag{3.22}$$

discontinuities in the second derivatives represent abrupt changes in heat capacity, isothermal compressibility and the volume coefficient of thermal expansion. Figure 3.7 illustrates the changes in heat capacity, as β-brass undergoes a second order order-disorder transition at the critical temperature, T_c (479°C).

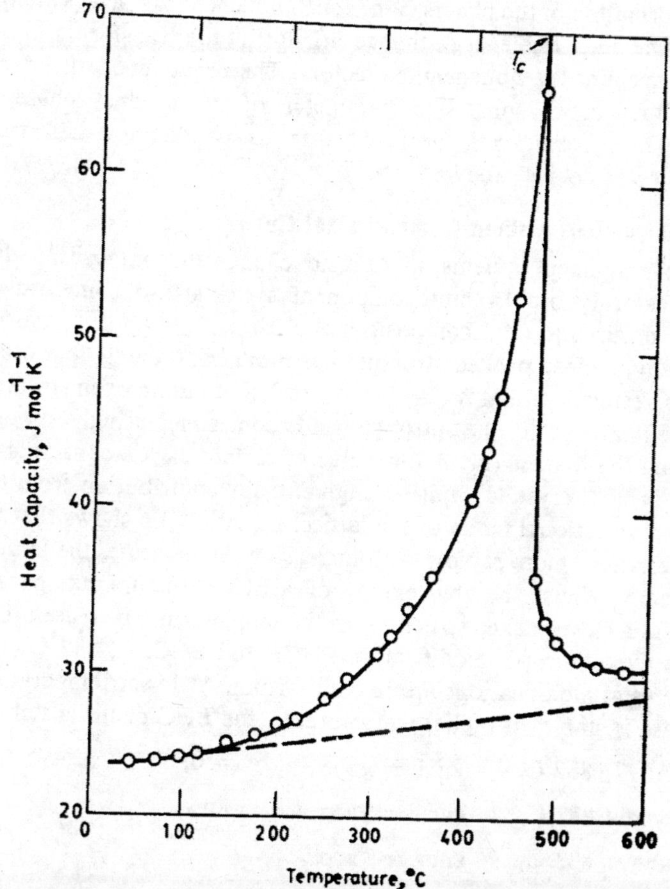

Fig. 3.7 Heat capacity of β-brass as a function of temperature through the order-disorder temperature (479°C).

3.4 DRIVING FORCE FOR FIRST ORDER TRANSFORMATIONS

A phase transformation that occurs spontaneously is accompanied by a decrease in the total free energy of the system. If g_β is the free energy per unit volume of the product phase β, and g_α is the corresponding free energy of the parent phase α, then during a spontaneous transformation at constant temperature and pressure

$$g_\beta - g_\alpha = \Delta g < 0 \tag{3.23}$$

The negative of the chemical free energy change $(-\Delta g)$ is regarded as the *driving force* for the phase change.

Even when the Gibbs free energy change is negative, there may be other non chemical factors which retard a phase transformation. Some of these factors are the creation of an interface between the parent and the product phases and the elastic strains resulting from the transformation. In a solid-solid transformation, the volume change during the transformation and/or any shape change needs to be accommodated either elastically or by the plastic deformation of the phases. When such nonchemical factors are present, we can define a net free energy change ΔG during the transformation, which takes into account the nonchemical factors. The negative of the derivative of ΔG with respect to some size parameter of the product phase can be called the *transformational force*. This is to be distinguished from the driving force as defined above.

Transformations without Compositional Change

In one component systems, there is no change in composition during a phase transformation. In multicomponent systems also, some transformations may occur without a compositional change.

Typically, a close packed structure has more bond energy per atom and a smaller residual enthalpy H_0 (or G) at 0 K, than an open structure. On raising the temperature, the entropy term becomes increasingly important in determining the free energy. A low value of certain elastic constants such as $c_{11} - c_{12}$ in BCC crystals implies a high entropy contribution from the corresponding vibrational mode of the lattice. Equation 3.6 shows that the free energy decreases more rapidly with increasing temperature, the larger is the heat capacity. Thus, the free energy of a BCC structure may become less than that of a close-packed structure, as the temperature increases. Titanium occurs in two allotropic modifications: HCP and BCC. At 882°C (1155 K), the two crystal modifications are in equilibrium. At lower temperatures, the HCP phase is stable; at higher temperatures, the BCC phase is stable.

At 882°C, $g_{HCP} - g_{BCC} = \Delta g = 0$,

below 882°C, $g_{HCP} - g_{BCC} < 0$, and

above 882°C, $g_{HCP} - g_{BCC} > 0$.

Other examples of such phase changes are those that occur in alkali metals. For example, sodium which is BCC at room temperature transforms to HCP structure at −237°C (36 K). However, this situation may not always be evident, if there are other contributions to the free energy. For example, the stable form of iron at low temperatures is the BCC form, as the free energy equations are here modified by magnetic contributions.

The application of hydrostatic pressure at constant temperature increases the enthalpy of a phase, since $H = E + PV$. This increase, if large enough, can raise the free energy of the existing phase sufficiently to give rise to a transformation from an open to a close packed structure. For example, the BCC iron (α) at room temperature transforms to a close packed HCP structure (ϵ) under a hydrostatic pressure of about 12 500 MPa at room temperature.

The differences in the free energies of elements in different crystal forms are called the lattice stability parameters. These data for a number of elements have been computed and compiled by Kaufman.

Transformations with Compositional Change

Consider a binary system of A and B. The free energies of two phases α and β in this system are shown schematically in Fig. 3.8 plotted against the concentration of B component at constant temperature and pressure. In the figure, the segment *ab* represents the chemical free energy change per mole for the transformation without a change in composition.

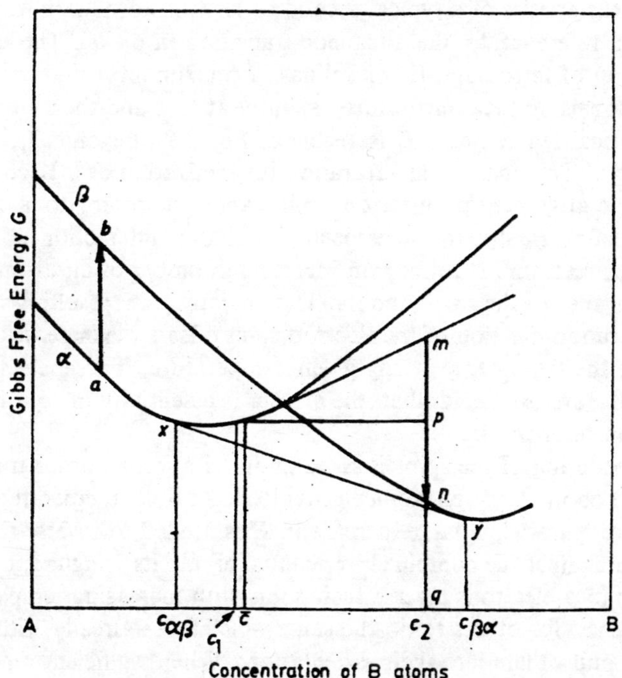

Fig. 3.8 Free energy G of the α and β phases as a function of composition in a binary system of A and B.

Now consider the formation of a small quantity of β out of the α phase of composition $\bar{c}$. Let the β phase have some general composition c_2, causing a small change in the composition of the matrix from $\bar{c}$ to c_1. Then it can be shown that the change in chemical free energy ΔG per mole of the β phase formed is

$$\Delta G = G^{\beta}_{c_2} - G^{\alpha}_{\bar{c}} - (c_2 - \bar{c}) \left.\frac{dG}{dc}\right|_{c=\bar{c}} \tag{3.24}$$

The magnitude of ΔG is equal to the segment mn in Fig. 3.8, drawn at c_2 between the tangent of the α free energy curve at $\bar{c}$ and the β free energy curve. If the β phase has the equilibrium composition $c_{\beta\alpha}$, the segment mn will be shifted to $c_{\beta\alpha}$. It may be noted that the composition of the β phase c_2 must be significantly different from the matrix composition $\bar{c}$ in order for a finite driving force to be available, i.e., the segment mn must be beyond the point of intersection of the β free energy curve with the tangent to the α curve at $\bar{c}$.

When we have a miscibility gap in a binary system, the free-energy/composition curve is continuous between the two coexisting phases, as shown in Fig. 3.5. Equation 3.24 is easily modified to be applicable here.

3.5 SECOND ORDER TRANSFORMATIONS

The characteristics of a typical second order transformation can be discussed with reference to the magnetic transition in nickel. The saturation magnetization of ferromagnetic nickel has a maximum value at 0 K and decreases on raising the temperature, slightly at first and then more rapidly until the critical temperature T_c is reached, Fig. 3.9. Beyond T_c, nickel is paramagnetic. The maximum saturation magnetization at 0 K corresponds to a complete alignment of electron spins and, therefore, to a minimum residual enthalpy H_0 due to the associated exchange interactions.

As the temperature is raised, an increasing number of electrons set their spins antiparallel to their neighbours. The entropy gain resulting from the progressive randomization of the electron spins offsets the increase in H so as to minimize the free energy at any given temperature. These electronic rearrangements are so rapid that the system is essentially in equilibrium on reaching that temperature.

The disordering of the spins is an example of a second order transformation and the cooperative phenomena involved. At low temperatures, when all spins are parallel, a large increase in H is needed to reverse the spin of one electron against the combined restraints of all its neighbours. But as more and more electrons reverse their spins with increasing temperature, it becomes easier for others to do the same, since those already with reversed spins try to pull others into their orientation. When sufficient numbers of spins point in both directions, the change in H with further disordering becomes negligible and general disorder sets in: this occurs at the transition temperature or the Curie point T_c. The cooperative action of the electron

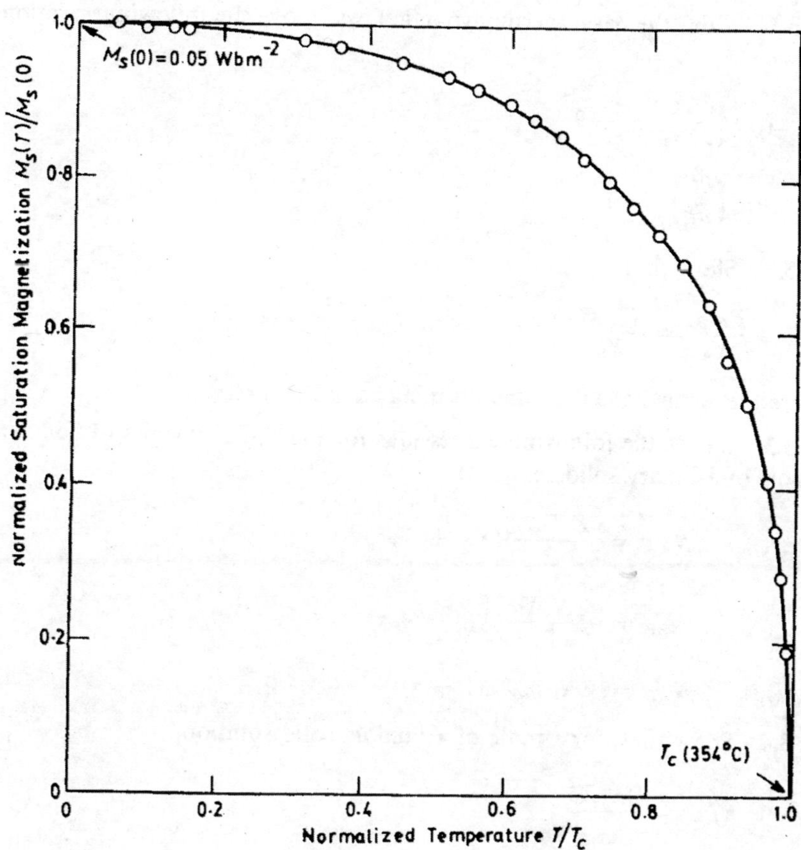

Fig. 3.9 Saturation magnetization of nickel as a function of temperature.

spins causes the saturation magnetization curve in Fig. 3.9 to drop off precipitously as the temperature approaches T_c.

Thus second order transformations occur over a range of temperature under equilibrium conditions. The associated excess heat capacity is correspondingly spread over a temperature range, as illustrated in Fig. 3.7. This behaviour is in contrast to the case of a first order transformation, where an entropy change occurs *at* the transformation temperature and the excess heat capacity becomes infinite at that temperature.

FURTHER READING

J.W. Christian, *The Theory of Transformations in Metals and Alloys*, Solid Solutions, Chapter 6, p. 169, Pergamon Press, Oxford (1975).

L. Kaufman and H. Bernstein, *Computer Calculations of Phase Diagrams*, Academic Press, New York (1970).

EXERCISES

3.1 From the basic thermodynamic laws, prove the following relationships:

(i) $\left(\frac{\partial G}{\partial T}\right)_P = -S$

(ii) $\left(\frac{\partial G}{\partial P}\right)_T = V$

3.2 Show that

$$\frac{P_{A(B)}}{X_A} = \frac{P_{B(A)}}{X_B}$$

where the terms have the same meaning as in the text.

3.3 Derive the following expressions for the different types of bonds in a mole of a binary solid solution:

(i) $N_{AA} = \frac{ZN_0X_A}{2}(X_A + X_B\alpha)$

(ii) $N_{BB} = \frac{ZN_0X_B}{2}(X_B + X_A\alpha)$

(iii) $N_{AB} = ZN_0X_AX_B(1 - \alpha)$.

3.4 Show that, for a mole of a random solid solution,

$$N_{AA} = \frac{ZN_0X_A^2}{2}$$

whereas for a mechanical mixture of pure A and pure B of the same overall composition

$$N_{AA} = \frac{ZN_0X_A}{2}$$

3.5 Show that the free energy of a mixture of two phases in equilibrium in a binary system is given by the point on the common tangent line that corresponds to that overall composition. Show that the free energy of a mixture of phases of any other composition or a single phase is higher.

3.6 Prove the relationship given in Eq. 3.24.

4

Nucleation Kinetics

In the introductory chapter, we referred to the category of transformations which are of the nucleation-and-growth type. In nucleation, tiny volumes of the product phase called nuclei, assumed to be identical in structure and composition to the product phase, form first. These subsequently grow into the matrix. The nucleation process occurring by thermal fluctuations is also known as *heterophase fluctuation* (embryo of another phase forming in the matrix phase out of thermal fluctuations).

In *random or homogeneous nucleation*, the probability of nucleation at any given site is identical to that at any other site within the assembly. In *nonrandom or heterogeneous nucleation*, the probability of nucleation at certain preferred sites in the assembly is much greater than that at other sites. In gases, nonrandom nucleation can occur at the container walls or at suspended impurity particles. In liquids, suspended solid particles and the walls of the container are preferred sites for nucleation. In solid-state transformations, homogeneous nucleation is the exception rather than the rule. Inclusions, grain boundaries, stacking faults, dislocations and possibly point imperfections act as preferred nucleation sites.

4.1 HOMOGENEOUS NUCLEATION

Volmer and Weber, and Becker and Doring developed a theory of nucleation for the formation of liquid droplets in supersaturated vapours. A basic assumption of this theory is that, by random thermal fluctuations, very small droplets of the product phase, identical in structure and composition to the final liquid phase, form in the supersaturated parent (vapour) phase. The small droplets, which may contain of the order of 100 atoms, are assumed to have the same surface and thermodynamic properties as the bulk liquid.

The assumptions of the Volmer-Weber theory are open to question at very small sizes of the nucleating particle. A particle with about 100 atoms would be $\sqrt[3]{100} = 4\text{-}5$ atomic diameters in size. The size is comparable in magnitude to the thickness of the transition zone (a zone of intermediate density between the liquid and the vapour). In such a situation, the division of the free energy into bulk (Gibbs) free energy and surface energy is quite arbitrary. Moreover, the surface energy of the interface will vary significantly with the curvature of the interface, which is very sharp for small particles of this size range. Nevertheless, these ideas have been carried over to solid-state nucleation.

The Homogeneous Nucleation Barrier

If ΔG is the free energy change accompanying the formation of a new particle of the transformation product, we can write:

$$\Delta G = V\Delta g + S_A\sigma \tag{4.1}$$

where V and S_A denote the volume and the surface area of the particle. Δg is the chemical free energy change per unit volume of the product phase. σ is the specific surface energy of the interface separating the product and the parent phases. When there is supersaturation in the parent phase, Δg is negative. The surface energy term in Eq. 4.1 is, however, positive. In the absence of strains, the particle tends to have a spherical shape to minimize its surface area. For a spherical particle of radius r,

$$\Delta G = \frac{4}{3}\pi r^3\Delta g + 4\pi r^2\sigma \tag{4.2}$$

This function passes through a maximum with increasing r whenever Δg is negative. Initially, as the new phase particle starts to form, the energy of the system increases, as the surface energy term is dominant. At the maximum, the variation with r of the surface energy term and the volume energy term exactly balance each other. Thereafter, the variation in the volume energy term becomes dominant. As this term is negative, there is a continuous decrease in the energy of the system beyond the maximum, as illustrated in Fig. 4.1. The temperature dependence of ΔG is also shown in Fig. 4.1. At $T = T_0$, $\Delta g = 0$, so that the ΔG-r curve increases continuously, without passing through a maximum. At $T < T_0$, $\Delta g < 0$, and the curve has a maximum. The maximum value of ΔG decreases with increasing driving force, i.e. with increasing magnitude of Δg. The temperature dependence of σ is usually small enough to be neglected. Setting $d\Delta G/dr = 0$, the following critical values (denoted by the asterisk) are obtained:

$$r^* = -\frac{2\sigma}{\Delta g} \tag{4.3}$$

$$\Delta G^* = \frac{16\pi\sigma^3}{3(\Delta g)^2} \tag{4.4}$$

These critical values are indicated in Fig. 4.1. ΔG^* is known as *the nucleation barrier*. Particles which are smaller than the critical size characteristic of a particular temperature are called *embryos*. Those which are larger than the critical size are called *nuclei*. The particle with a radius equal to r^* is a *critical nucleus*. As Δg becomes more negative with a lowering of the temperature, the critical condition occurs at smaller values of ΔG^* and r^*, see Fig. 4.1.

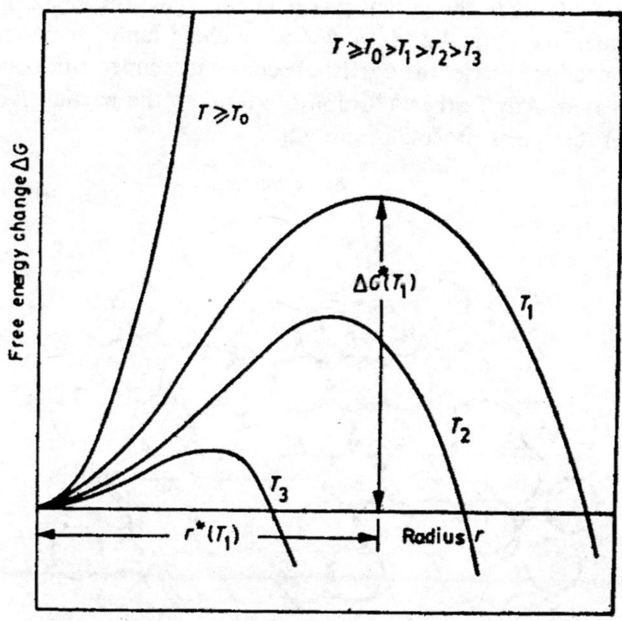

Fig. 4.1 The free energy change ΔG during nucleation as a function of the radius r of a spherical particle at different temperatures.

As data on the heat of reaction (the enthalpy change during the transformation) are more readily available than on the Gibbs free energy change, the following simplification is often used:

$$\Delta g \approx \frac{\Delta h(T_0 - T)}{T_0}$$

$$= \frac{\Delta h \, \Delta T}{T_0} \tag{4.5}$$

where T is the transformation temperature and Δh is the enthalpy change per unit volume of the product formed, taken to be independent of the temperature of the transformation. ΔT is the degree of undercooling or supercooling. Combining Eqs. 4.4 and 4.5, we obtain

$$\Delta G^* = \frac{16\pi\sigma^3 T_0^2}{3(\Delta h)^2(\Delta T)^2} \tag{4.6}$$

Equation 4.6 indicates that the nucleation barrier varies inversely as the square of the degree of supercooling.

Rate of Homogeneous Nucleation

In the classical theory, the process of nucleation is defined as the addition of one atom to a critical-sized particle such that it becomes *just supercritical.*

Let there be s^* atoms in the parent phase facing the critical-sized particle across the interface, Fig. 4.2. If any one of these jumps from the parent phase to the product phase, the particle becomes just supercritical and is said to have nucleated. Any further addition of atoms to the particle is considered to be the subsequent process of growth.

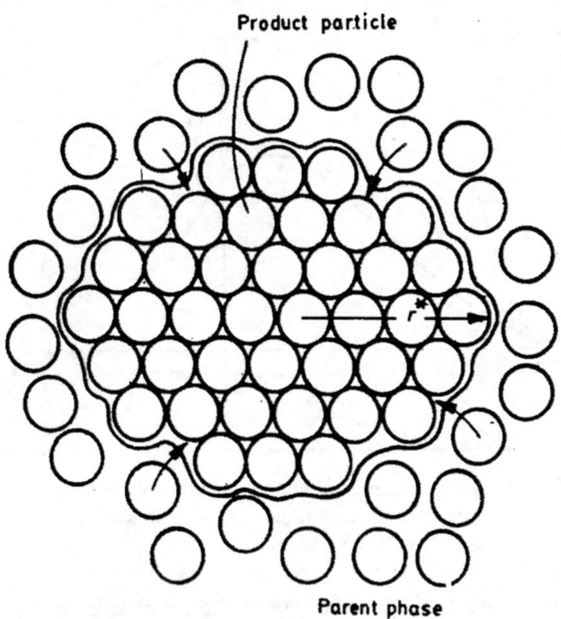

Fig. 4.2 The transfer of atoms across the interface from the parent phase to the product particle of critical radius r^*.

The *Maxwell-Boltzmann* statistics can be used to estimate the number of critical-sized particles in the parent phase. Let the total number of particles per unit volume of the parent phase be N_T. Then the number of critical-sized particles N^* is approximated by

$$N^* = N_T \exp\left(-\frac{\Delta G^*}{kT}\right) \tag{4.7}$$

where ΔG^* is the nucleation barrier. The frequency ν' with which the s^* atoms surrounding the critical-sized particle can cross the interface to join the particle is given by

$$\nu' = s^*\nu \exp\left(-\frac{\Delta G_D}{kT}\right) \tag{4.8}$$

where ν is the lattice vibration frequency (which is $\sim 10^{13}\ \mathrm{s^{-1}}$ above the Debye temperature) and ΔG_D is the activation barrier for diffusion across the interface. Then the rate of homogeneous nucleation $I(=\mathrm{d}N/\mathrm{d}t)$ is the number of nucleation events per unit volume per unit time:

$$I = N^*\nu'$$

$$= N_T \nu s^* \exp\left[-\frac{(\Delta G^* + \Delta G_D)}{kT}\right] \tag{4.9}$$

As a first approximation, in homogeneous nucleation, N_T can be taken to be the number of atoms in the parent phase per unit volume ($\sim 10^{29}$ m^{-3}). Then the pre-exponential term $N_T \nu s^*$ is 10^{43} m^{-3} s^{-1} in order of magnitude. A measurable nucleation rate is about 10^6 m^{-3} s^{-1}. Then, for a measurable rate, the exponent in Eq. 4.9 must be about 85. The nucleation rate is very sensitive to the value of this exponent; if the latter is decreased to about 70, say, by increasing the driving force $(-\Delta g)$ through a change in the reaction temperature, I can be increased by a factor of 10^6, i.e., to 10^{12} m^{-3} s^{-1}.

The Temperature Dependence of the Nucleation Rate

As already noted in Eq. 4.6, ΔG^* is a function of temperature. Its value at the equilibrium temperature T_0, where $\Delta T = 0$, is infinite. Therefore, the nucleation rate I as given by Eq. 4.9 is zero at the equilibrium temperature. At temperatures below T_0, $\Delta T > 0$ and ΔG^* is finite. As ΔG^* decreases with decreasing temperature, I increases. At some degree of supercooling, I reaches a maximum value. Thereafter, the decrease in the thermal energy kT dominates through the exponent and the nucleation rate decreases with decreasing temperature. It becomes zero at 0 K.

The temperature of the maximum rate of nucleation, $T_{I_{max}}$, can be found by setting $dI/dT = 0$, from which we obtain the following expression for $T_{I_{max}}$

$$\frac{d\Delta G^*}{dT} = \frac{(\Delta G^* + \Delta G_D)}{T_{I_{max}}} \tag{4.10}$$

Figure 4.3 illustrates a graphical method of determination of $T_{I_{max}}$, when ΔG^* is known as a function of temperature.

The Time Dependence of the Nucleation Rate

The nucleation rate given in Eq. 4.9 is time independent. It assumes that N^* critical-sized particles are always available, as given by the equilibrium statistical distribution. This may not be the case, especially in the initial stages after quenching a material from a higher temperature T' to the reaction temperature T_1. If $T' > T_0 > T_1$, initially an incubation period may be present, when the nucleation rate is negligible. After some time, the number of critical-sized particles N^* builds up to the steady-state value characteristic of T_1. During this period, I is expected to increase with time according to the relation

$$\frac{I_t - I_i}{I_s - I_i} = \exp\left(-\frac{\omega}{t}\right) \tag{4.11}$$

where I_t is the nucleation rate at any time t, I_i is the initial nucleation rate,

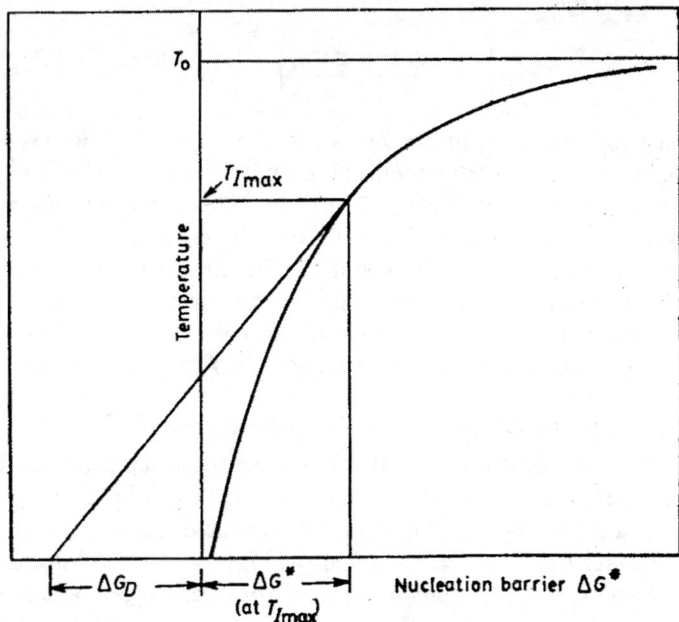

Fig. 4.3 Graphical determination of the temperature of the maximum rate of nucleation, $T_{I\max}$.

I_s is the nucleation rate when the steady state is attained, and ω is the time constant. This variation in the nucleation rate is illustrated in Fig. 4.4. At $t = 0$, $I = I_i \sim 0$. At large values of t, $I \to I_s$. The value of the time constant ω determines whether the change in the nucleation rate will be transient or prolonged. If ω is small, the effect will be transient. If ω is large, the transformation may be complete, even before the steady state nucleation rate is attained.

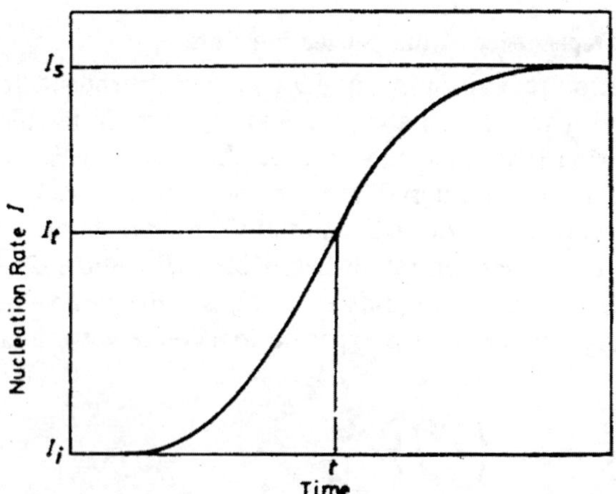

Fig. 4.4 The time dependence of the nucleation rate I.

4.2 HOMOGENEOUS NUCLEATION WITH COMPOSITIONAL CHANGE

Becker's Model

Homogeneous nucleation with a compositional change in systems having two or more components has been considered by several authors. Becker used an expression similar to Eq. 3.24 for the chemical free change during the transformation when there is an attendant compositional change. The nucleating particle is assumed to have a uniform and fixed composition corresponding to that of the stable phase. The crystal lattices of the parent and the product phases are assumed to have perfect matching at the interface. The surface energy here is solely due to the abrupt compositional change at the interface. It is evaluated on the basis of the nearest neighbour bond energies and turns out to be proportional to the square of the difference in composition across the interface. Using these fixed values of the chemical free energy change (at a constant temperature) and the surface energy, Becker calculated the critical nucleation barrier and the critical radius of the particle, using Eqs. 4.3 and 4.4. The temperature of the maximum rate of nucleation calculated for a Au-30 at.% Pt alloy using an expression similar to Eq. 4.10 gave 900°K, which compared well with the temperature of 860 K corresponding to the minimum time measured for half-transformation in that alloy.

Borelius' Model

Borelius considered the alternative possibility of a continuous variation in composition of a nucleus of a given size. Referring to Fig. 4.5, the segment *mn* represents the free energy change when a nucleating particle has a composition to the right of the matrix composition $\bar{c}$. The free energy change is initially positive and increases as the composition shifts more to the right. It reaches a maximum value $m'n'$ at the point of tangency of the second tangent to the free energy curve drawn parallel to the tangent at the matrix composition $\bar{c}$, see Fig. 4.5. Neglecting the energy of the interface, this is the free energy barrier to nucleation with the composition axis as the reaction path. The free energy barrier decreases, as the composition of the matrix shifts still further to the right and becomes zero at the point of intersection q of the tangent at $\bar{c}$ with the free energy curve. The nucleation barrier disappears at and beyond this point.

Hobstetter's Model

Hobstetter allowed both the nucleus composition and the radius of the spherical nucleating particle to vary simultaneously. The surface energy was evaluated following Becker's approach described above. The critical nucleation barrier corresponds to the *saddle point* on the free energy surface plotted with composition and radius as the two axes. There is a critical composition c^* and a critical radius r^* corresponding to the saddle point.

Even though the Hobstetter model is an improvement over the earlier

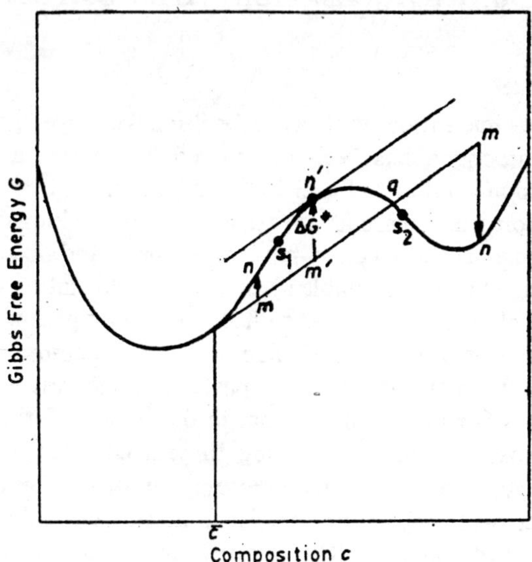

Fig. 4.5 The free energy change during nucleation of a particle of composition different from the matrix composition $\bar{c}$.

models of Becker and Borelius, it also suffers from the arbitrary assumption of a sharp interface and the neglect of the strain energy that will arise from such a coherent compositional fluctuation. It remained for Cahn and his associates to generalize the treatment by dropping the assumption of a sharp interface. The gradient surface energy arising out of a diffuse interface and the strain energy were taken into account. This treatment, known as *the spinodal decomposition,* is valid for compositions that lie within the two inflection points of the continuous free energy-composition curve and is described in Chapter 12.

4.3 STRAIN ENERGY EFFECTS

A change in volume accompanies most phase transformations. When the adjoining parent phase is a fluid, these volume changes can be accommodated easily by flow in the parent phase. If, however, the reaction occurs entirely within the solid state, the volume change introduces strains in the phases and the nucleation process will be inhibited due to the associated strain energy. In martensitic transformations, besides any change in volume, the transforming region also undergoes a change of shape, resulting in additional strains.

The nature of the interface between the parent phase and the product particle is important in determining the strain energy. So, the types of possible interfaces and their structure must be understood, before estimates of the strain energy for different interfaces can be made.

The Structure of the Interface

The structure of the interface between the parent and the product phases is determined by the degree of matching between the two crystals across the interface. Interfaces are classified into three groups depending on the degree of matching.

When there is one-to-one lattice registry along the interface, as illustrated in Fig. 4.6(a), the interface is said to be *coherent*. Note that the interface lies perpendicular to the plane of the paper and the lattice registry is required also in the direction that extends into the paper. Interatomic distances along two directions of the parent and the product crystals will rarely be identical, but may differ only slightly from each other. Under such conditions, coherency may still prevail, but some elastic strains will be introduced due to the forced matching at the interface. Coherent interfaces usually have a low surface energy, in the range of 0.01-0.05 J m^{-2} (10-50 erg/cm^2). If the crystal structures of the parent and the product phases happen to match well across the interface, the nucleation of the product phase will be greatly facilitated as a result of the low interfacial energy, recall from Eq. 4.4 that the nucleation barrier is proportional to the cube of the interfacial energy.

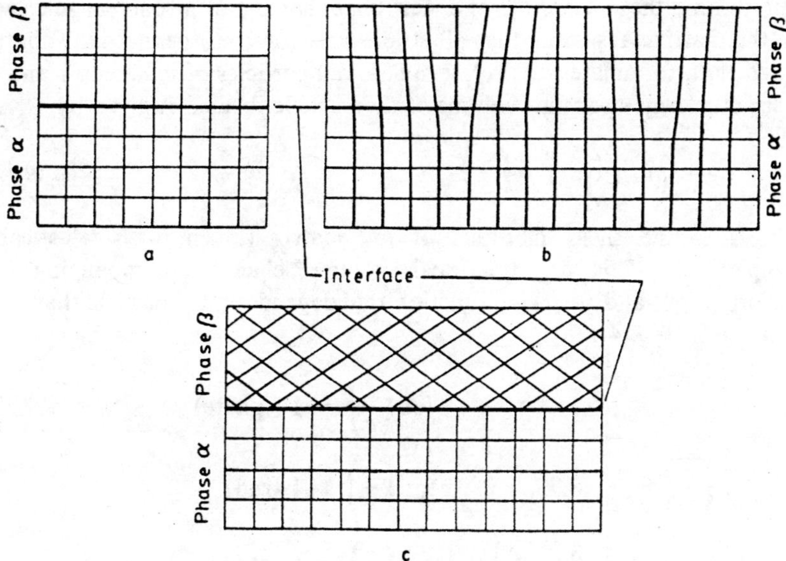

Fig. 4.6 (a) Coherent, (b) semicoherent, and (c) incoherent interfaces.

When the mismatch between the parent and the product phases is appreciable, the associated elastic strains due to coherency may be excessively large. In such instances, regions of good coherency in the interface can be separated by localized regions of severe distortion comprising of interfacial dislocations, as illustrated in Fig. 4.6(b). Here, along the interface, the spacing between seven atomic planes in the α phase is seen to be equal to the spacing between eight planes in the β phase. Accordingly, every eighth plane in the β phase forms an edge dislocation at the interface, allowing one-to-one registry of the intervening seven planes. A similar description will be valid for the misfit in the direction perpendicular to the plane of the paper, the corresponding array of interfacial dislocations lying in that direction. Such an interface is said to be *semicoherent.* Its energy is largely composed of the elastic and core energies of the interfacial dislocations. In estimating the elastic energy of an interfacial dislocation, its elastic strain field can be taken to extend on either side up to half of the distance of separation between neighbouring dislocations. The surface energy of a semicoherent interface lies in the range of 0.1-0.3 J m^{-2} (100-300 erg/cm^2).

When the two crystal structures are very different from each other, relatively little matching is possible and a highly distorted *incoherent interface* results, as shown schematically in Fig. 4.6(c). The energy of this interface may be in the range of 0.4-1.0 J m^{-2} (400-1000 erg/cm^2).

Strain Energy due to Volume Change

Nabarro considered the uniform expansion or contraction which may occur during the formation of a new phase having an incoherent interface with the matrix. Assuming that all strains are stored in the matrix, Nabarro showed that, for oblate and prolate spheroidal particles of semi-axes r and c, the strain energy ϵ per unit volume of the particle formed is given by

$$\epsilon = \frac{2}{3}\mu_m\left(\frac{\Delta V}{V}\right)^2 \phi\left(\frac{c}{r}\right) \tag{4.12}$$

where μ_m is the shear modulus of the matrix (taken to be elastically isotropic), $\Delta V/V$ is the fractional volume change accompanying the transformation, and $\phi(c/r)$ is a function that depends on the particle shape, as illustrated in Fig. 4.7.

$$\phi\left(\frac{c}{r}\right) = \begin{cases} 1 & \text{for } \left(\frac{c}{r}\right) = 1 \text{ (sphere)} \\ 0.75 & \text{for } \left(\frac{c}{r}\right) \gg 1 \text{ (rod)} \\ \frac{3}{4}\pi\left(\frac{c}{r}\right) & \text{for } \left(\frac{c}{r}\right) \ll 1 \text{ (disc)} \end{cases}$$

Equation 4.12 can be written as

$$\epsilon = A_{\text{dil}}\,\phi\left(\frac{c}{r}\right) \tag{4.13}$$

where A_{dil} is the strain energy factor due to dilatational strains. For a disc with $(c/r) = 1/20$ and for a volume change of 5%, ϵ has a value of $\sim$15 MJ m^{-3} for typical solid-state transformations in metallic systems.

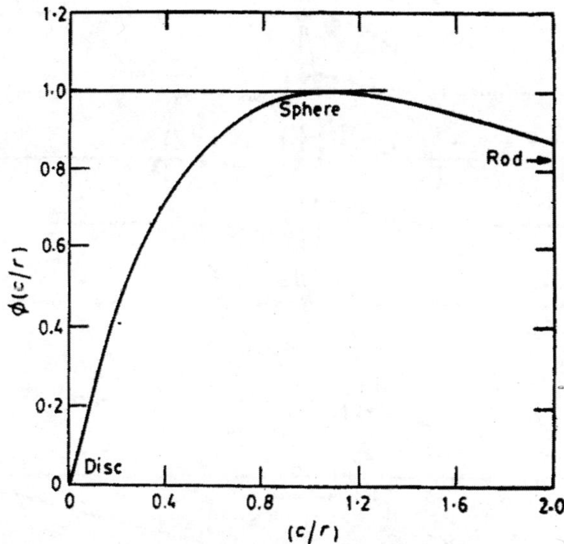

Fig. 4.7 The strain energy parameter $\phi(c/r)$ as a function of (c/r) in Nabarro's model.

Strain Energy due to Shape Change

Fisher, Hollomon and Turnbull have estimated the shear strain energy resulting from a change of shape of a transforming region. Such a change of shape occurs in martensitic transformations. Let a plate-like particle of the product phase form in the parent matrix. Parallel fiducial lines drawn on the prepolished matrix get tilted due to the shape change, as indicated in Fig. 4.8. All the strains due to the shape change are assumed to reside in the matrix within a sphere circumscribed about the disc-shaped particle as shown in Fig. 4.8(a). It is assumed that the strain is uniformly distributed within the sphere. Referring to Fig. 4.8(b), if ϕ is the shear angle of the transformation,

$$\tan \phi = \frac{\delta}{c} \tag{4.14}$$

The shear strain within the sphere is

$$\tan \psi \approx \frac{\delta}{r} \tag{4.15}$$

Substituting for δ from Eq. 4.14 into 4.15, we have

$$\tan \psi = \tan \phi \frac{c}{r} \tag{4.16}$$

The shear strain energy per particle stored in the spherical volume is given by

$$\epsilon_p = \frac{4}{3} \pi r^3 \tfrac{1}{2}\mu_m \tan^2 \phi \frac{c^2}{r^2}.$$

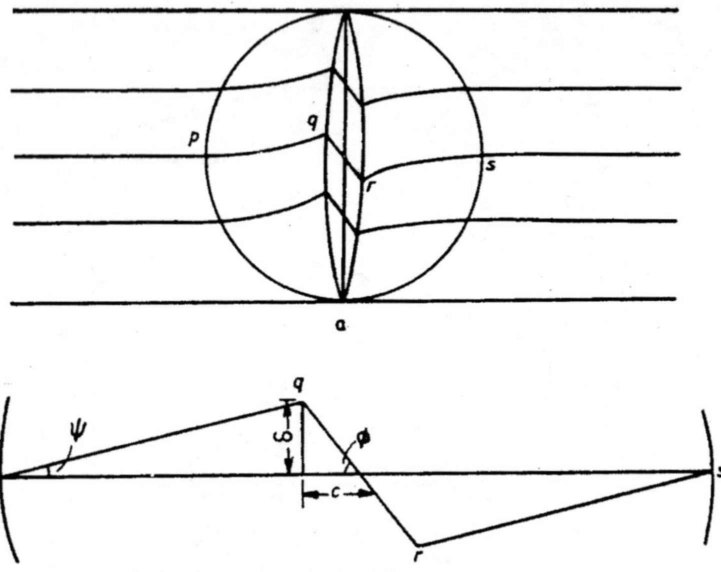

Fig. 4.8 Tilting of the fiducial lines when a transforming region changes shape.

The shear strain energy per unit volume of the particle is

$$\epsilon = \tfrac{1}{2}\mu_m \tan^2 \phi \frac{c}{r}$$

$$= A_{\text{shear}} \frac{c}{r} \tag{4.17}$$

where $A_{\text{shear}} = \frac{1}{2}\mu_m \tan^2 \phi$. (The volume of the disc-shaped particle is $4/3\pi r^2 c$.) A typical value of the shear strain in martensitic transformations in steels is 0.2. Using $\mu_m = 80$ GN m^{-2}, A_{shear} is in the range of 1600 MJ m^{-3} and for $c/r = 1/20$, ϵ has a value of 80 MJ m^{-3}.

Strain Energy due to Coherency

Strain energy may also arise due to coherency when the crystal structures of the product phase and the surrounding matrix are constrained to match at the interface. As already pointed out on page 55, moderate mismatches between the two structures can elastically strain the phases or accommodate themselves in the form of interfacial dislocations. Consider the case where the mismatch results in both elastic coherency strains and misfit dislocations.

Let $\overset{0}{a}_m > \overset{0}{a}_p$, where a is the lattice constant or atom spacing along a direction of matching. m and p stand for matrix and product particle, respectively. The superscript ^{0} refers to the relaxed or natural (unstrained) condition. The natural misfit δ^0 is then defined as

$$\delta^0 = \frac{\overset{0}{a}_m - \overset{0}{a}_p}{\overset{0}{a}_m} \tag{4.18}$$

The actual misfit δ during the nucleation event is

$$\delta = \frac{a_m - a_p}{a_m} \tag{4.19}$$

Then the coherency strain in the matrix is

$$\epsilon_m = \frac{a_m - \overset{0}{a}_m}{\overset{0}{a}_m} \tag{4.20}$$

We can similarly define the coherency strain in the particle. The resulting strain energy per unit volume of the particle depends on the relative stiffness of the matrix and the particle. Assuming isotropic elasticity, Eshelby has estimated the strain energy for different situations. His analysis is based on the following conceptual steps:

1 Cut out the region which is to be transformed into the product

2 Allow the transformation of the region freely

3 Apply surface tractions (forces) to restore the region to its original form

4 Put back the region and weld the surfaces

5 Apply a layer of body force at the interface to annul the tractions applied in (3). This force will strain the matrix and the particle.

Eshelby's results yield the strain energy ϵ per unit volume of the product phase as given below for different cases.

Let Y_m and Y_p be the Young's moduli of the matrix and the particle and ν_m and ν_p be the corresponding Poisson's ratios. For equal stiffness ($Y_m = Y_p$), the strain energy is stored in equal amounts in the matrix and the particle:

$$\epsilon = \frac{Y\epsilon^2}{1 - \nu} \tag{4.21}$$

For a rigid matrix ($Y_m \gg Y_p$), all the strains are stored in the particle, as it is energetically favourable to be so:

$$\epsilon = \frac{3Y_p\epsilon_p^2}{2(1 - \nu_p)} \tag{4.22}$$

For a rigid particle ($Y_m \ll Y_p$), all the strains reside in the matrix:

$$\epsilon = \frac{3Y_m\epsilon_m^2}{(1 + \nu_m)} \tag{4.23}$$

Eshelby's model predicts that the coherency strain energy is *independent of particle shape*, in contrast to the results of Nabarro (Eq. 4.12) and Fisher, Hollomon and Turnbull (Eq. 4.17).

A 5% volume change during a transformation corresponds to a misfit of about 1.7% in a direction lying on the interface. For $Y_m = 200$ GN m^{-2}

and $\nu_m = 0.25$, the strain energy per unit volume of the particle is about 140 MJ m^{-3}. Thus the strain energy during coherent nucleation can be quite large.

Nucleation Barrier with Strain Energy

The equations for the free energy change ΔG during nucleation must be modified for solid-state transformations to take the strain energy into account. If the strain energy is independent of particle shape as in coherent nucleation, the particle can form as a sphere. The equations for r^* and ΔG^* are then modified as under:

$$r^* = -\frac{2\sigma}{(\Delta g + \epsilon)} \tag{4.24}$$

$$\Delta G^* = \frac{16\pi\sigma^3}{3(\Delta g + \epsilon)^2} \tag{4.25}$$

Under such conditions, nucleation becomes feasible only when $|\Delta g| > \epsilon$.

If the strain energy is a function of the particle shape, the balance between the interfacial energy and the strain energy results in a disc-shaped particle of semithickness c and radius r. With $c \ll r$, the volume V of such a particle is

$$V = \frac{4}{3}\pi r^2 c \tag{4.26}$$

The surface area S_A of the particle can be approximated to

$$S_A = 2\pi r^2 \tag{4.27}$$

If Eqs. 4.12 and 4.17 are applicable for estimating the strain energy, we can write

$$\Delta G = \frac{4}{3}\pi r^2 c\Delta g + 2\pi r^2\sigma + \frac{4}{3}\pi r^2 cA\left(\frac{c}{r}\right) \tag{4.28}$$

where $A = A_{dil} + A_{shear}$. The free energy of nucleation is now a function of two particle parameters describing the particle size: c and r. The plot of ΔG against these two parameters yields a surface. The free energy barrier will now be at the *saddle point* of the free energy surface, as illustrated in Fig. 4.9.

By setting $\partial(\Delta G)/\partial c = 0$ and $\partial(\Delta G)/\partial r = 0$, the critical values of the parameters denoted by the superscript * are obtained:

$$c^* = -\frac{2\sigma}{\Delta g} \tag{4.29}$$

$$r^* = \frac{4A\sigma}{(\Delta g)^2} \tag{4.30}$$

$$\Delta G^* = \frac{32\pi A^2\sigma^3}{3(\Delta g)^4} \tag{4.31}$$

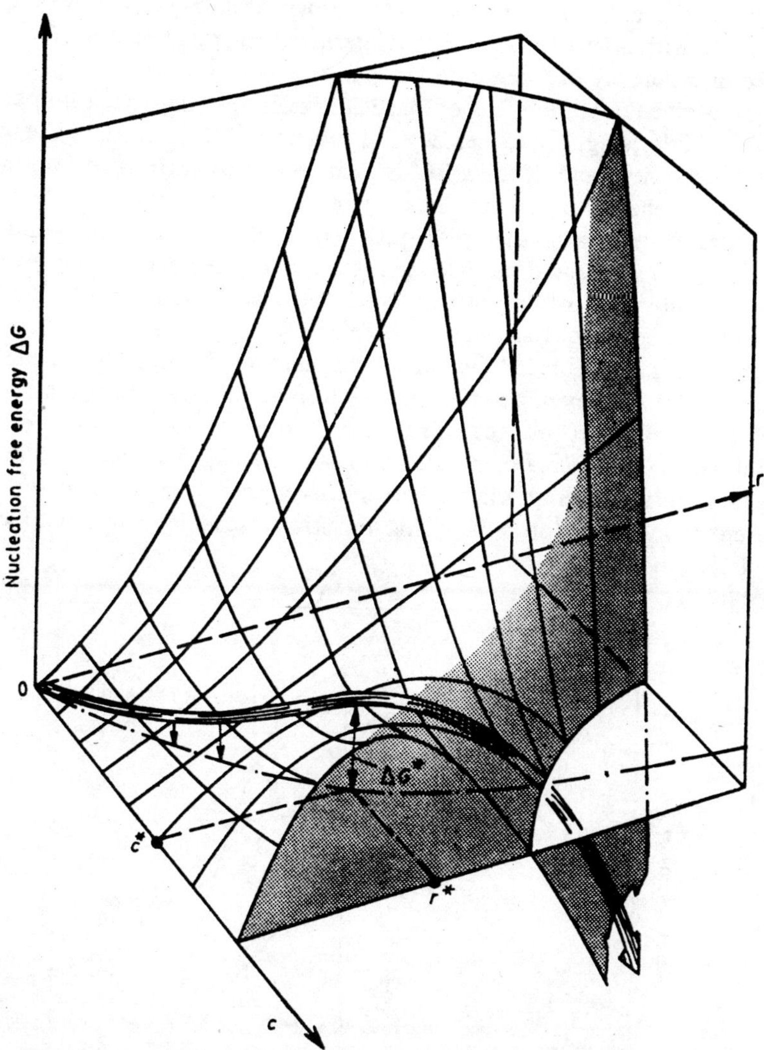

Fig. 4.9 The free energy surface with a saddle point, when ΔG is plotted against the radius r and the semithickness c of a disc-shaped particle.

$$V^* = -\frac{128\pi A^2\sigma^3}{3(\Delta g)^5} \tag{4.32}$$

When the misfit at the interface is accommodated both as elastic coherency strains and misfit dislocations, the expression for ΔG will include an additional variable δ, which is the optimum misfit that corresponds to the minimum total energy due to the coherency strains and misfit dislocations. If δ^0 in Eq. 4.18 happens to be large, nucleation tends to occur without much coherency, because the associated coherency strain energy will then be excessive. Under such conditions, ΔG^* is minimized mainly with misfit dislocations. If δ^0 is small, nucleation tends to occur with a high degree of coherency, inasmuch the coherency strain energy is then relatively low. Here, ΔG^* is minimized with a small value of the interfacial energy because of a high degree of coherency that can be tolerated.

For a given δ^0, the effect of particle size is also important in that the strain energy arising from coherency is a function of the volume of the particle. On the other hand, if the misfit is accommodated entirely by interfacial dislocations, the energy due to this contribution is proportional to the interfacial area. As we have noted previously, the surface area term dominates the volume term at small particle sizes, but the reverse is true at large particle sizes. This situation may result in coherent nucleation of a particle, but as it grows, the coherency strain energy becomes so large that it is no longer energetically favourable for the particle to remain coherent. The system can lower the free energy more effectively by forming interfacial dislocations to take up the misfit between the two phases. In other words, beyond a certain particle size, the decrease in coherency strain energy is greater than the increase in surface energy, when the interface changes from coherent to semi-coherent or incoherent structure. This transition is schematically illustrated in Fig. 4.10.

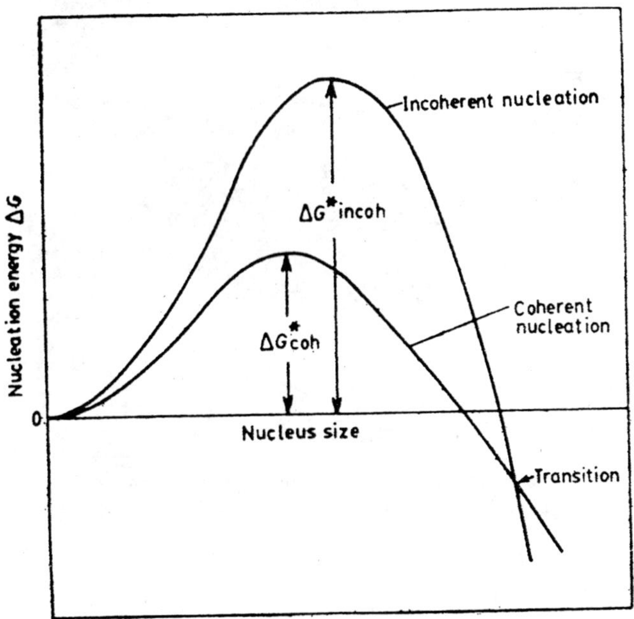

Fig. 4.10 Schematic illustration on the transition from a coherent to an incoherent interface, as the size of the nucleating particle increases.

4.4 HETEROGENEOUS NUCLEATION

As already indicated, nucleation in the solid state is quite often heterogeneous or nonrandom. Among the preferred nucleation sites are grain boundaries, grain edges and grain corners, dislocations and surfaces of inclusions or precipitates embedded in the matrix. We shall review here the equations which describe the nucleation kinetics for the process taking place at such special sites.

Nucleation on Grain Boundaries

Clem and Fisher have developed a model for the nucleation kinetics on grain boundaries, based on the following assumptions:

1 Any special orientation relationships between the nucleating particle and the matrix are neglected

2 The interface between the particle and the matrix is incoherent

3 The strain energy due to any volume change is neglected

4 During nucleation, surface tension forces due to the interfaces involved are in equilibrium.

In Fig. 4.11, the nucleation of the new phase particle at the grain boundary between two α grains is schematically illustrated. The β particle is lens-shaped in the form of *two* spherical caps. The circular plane of intersection of the cap lies in the grain boundary plane. The included lens angle 2θ (see Fig. 4.11) is determined by the equilibrium between the horizontal components of the surface tensions $\sigma_{\alpha\alpha}$ and $\sigma_{\alpha\beta}$:

$$\sigma_{\alpha\alpha} = 2\sigma_{\alpha\beta} \cos \theta \tag{4.33}$$

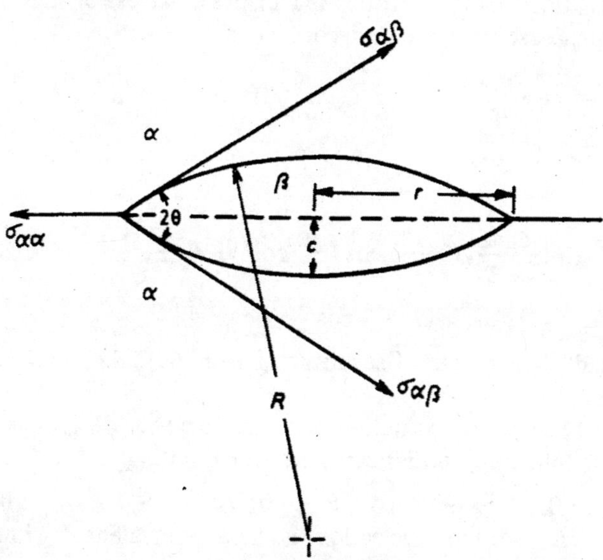

Fig. 4.11 Nucleation of a β particle in the shape of a full lens at an α-α grain boundary.

Note that the surface tension has the same dimensions ($N\ m^{-1}$) as surface energy ($J\ m^{-2}$), as $J = N\ m$.

Let R be the radius of curvature of the spherical caps and c the semithickness of the double lens. r is the radius of the circle of intersection. The following results hold good.

$$r = R \sin \theta \tag{4.34}$$

$$c = R(1 - \cos \theta) \tag{4.35}$$

The volume of the particle V can be shown to be

$$V = \frac{2}{3} \pi R^3 (2 - 3 \cos \theta + \cos^3 \theta) \tag{4.36}$$

and the surface area of the particle S_A is

$$S_A = 4\pi R^2 (1 - \cos \theta) \tag{4.37}$$

The free energy change during this type of nonrandom nucleation should now take into account the *energy gained* in the elimination of grain boundary area equal to πr^2:

$$\Delta G_{\text{het}} = \frac{2}{3} \pi R^3 (2 - 3 \cos \theta + \cos^3 \theta) \Delta g + 4\pi R^2 (1 - \cos \theta) \sigma_{\alpha\beta} - \pi r^2 \sigma_{\alpha\alpha} \tag{4.38}$$

Substituting $r = R \sin \theta$ and $\sigma_{\alpha\alpha} = 2\sigma_{\alpha\beta} \cos \theta$ in Eq. 4.38, we obtain

$$\Delta G_{\text{het}} = \left(\frac{2}{3} \pi R^3 \Delta g + 2\pi R^2 \sigma_{\alpha\beta}\right)(2 - 3 \cos \theta + \cos^3 \theta) \tag{4.39}$$

Setting $d(\Delta G_{\text{het}})/dR = 0$, the following critical values of the parameters indicated by superscript * are derived:

$$R^* = -\frac{2\sigma_{\alpha\beta}}{\Delta g} \tag{4.40}$$

$$r^* = -\frac{2\sigma_{\alpha\beta} \sin \theta}{\Delta g} \tag{4.41}$$

$$\Delta G^*_{\text{het}} = \frac{8\pi\sigma^3_{\alpha\beta}}{3(\Delta g)^2} (2 - 3 \cos \theta + \cos^3 \theta) \tag{4.42}$$

$$= \tfrac{1}{2} \Delta G^*_{\text{homo}} (2 - 3 \cos \theta + \cos^3 \theta) \tag{4.43}$$

the last equality arising out of substituting from Eq. 4.4 for homogeneous nucleation.

Figure 4.12 shows the variation of the ratio $\Delta G^*_{\text{het}}/\Delta G^*_{\text{homo}}$ as a function of cos θ. The following special cases are worth noting:

1 $\sigma_{\alpha\alpha} \to 0$, $\theta \to 90°$, $\cos \theta \to 0$, $\Delta G^*_{\text{het}} \to \Delta G^*_{\text{homo}}$. The lens becomes a sphere, so the nucleation is here equivalent to homogeneous nucleation.

2 $\theta \to 0$, $\cos \theta \to 1$, $\Delta G^*_{\text{het}} \to 0$. The lens here degenerates into a vanishingly thin disc that spreads along the interface without any nucleation barrier. Here, $\sigma_{\alpha\alpha} \geqslant 2\sigma_{\alpha\beta}$. It is easily seen from Eqs. 4.36, 4.40 and 4.42, that

$$\Delta G^*_{\text{het}} = -\frac{\Delta g V^*}{2} \tag{4.44}$$

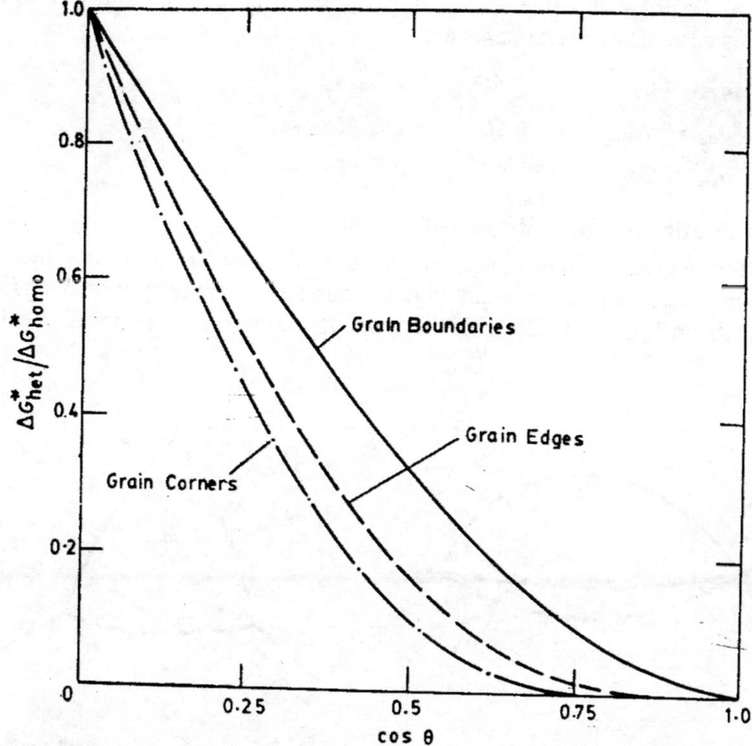

Fig. 4.12 The ratio $\Delta G^*_{het}/\Delta G^*_{homo}$ as a function of $\cos\theta$ in grain-boundary, grain-edge and grain-corner nucleation.

Nucleation on Grain Boundaries (Strain Energy Present)

When strain energy is taken into account during grain boundary nucleation, we drop the assumption that the surface tension forces are in equilibrium. Instead of the lens shape, we take the particle to be an oblate spheroid with volume $V = 4/3\ \pi r^2 c$ and surface area $S_A = 2\pi r^2$. Then the free energy of nucleation is given by

$$\Delta G_{het} = \frac{4}{3}\pi r^2 c\Delta g + 2\pi r^2 \sigma_{\alpha\beta} - \pi r^2 \sigma_{\alpha\alpha} + \frac{4}{3}\pi r c^2 A \tag{4.45}$$

where A is the strain energy parameter.

The critical values corresponding to the saddle point of the free energy surface are:

$$c^* = -\frac{2(\sigma_{\alpha\beta} - \frac{1}{2}\sigma_{\alpha\alpha})}{\Delta g} \tag{4.46}$$

$$r^* = \frac{4A(\sigma_{\alpha\beta} - \frac{1}{2}\sigma_{\alpha\alpha})}{(\Delta g)^2} \tag{4.47}$$

$$\Delta G^*_{het} = \frac{32\pi A^2(\sigma_{\alpha\beta} - \frac{1}{2}\sigma_{\alpha\alpha})^3}{3(\Delta g)^4} \tag{4.48}$$

If $\sigma_{\alpha\alpha} \to 0$, Eq. 4.48 becomes Eq. 4.31, previously derived for the homogeneous case. The special cases are:

1 $\sigma_{\alpha\alpha} \to 0$, $\quad \Delta G^*_{\text{het}} \to \Delta G^*_{\text{homo}}$

2 $\sigma_{\alpha\alpha} = \sigma_{\alpha\beta}$, $\quad \Delta G^*_{\text{het}} = 1/8\ \Delta G^*_{\text{homo}}$

3 $\sigma_{\alpha\alpha} \geqslant 2\sigma_{\alpha\beta}$, $\quad \Delta G^*_{\text{het}} = 0$.

Nucleation on Grain Edges and Corners

Grain edges are lines along which three neighbouring grains meet. A particle nucleating at a grain edge is bounded by three spherical surfaces, as shown in Fig. 4.13. Grain corners are points where four grains meet. A

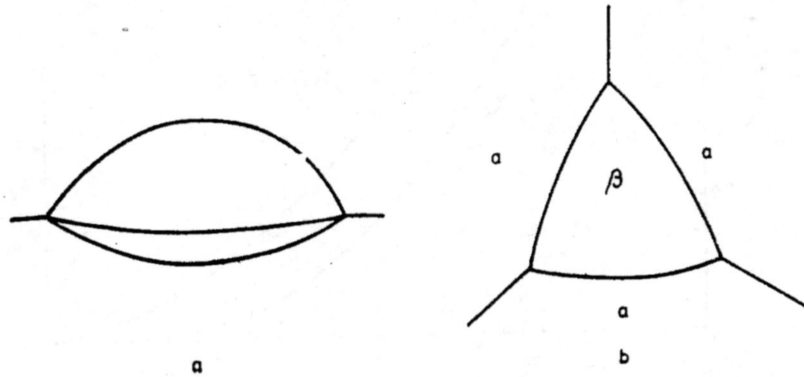

Fig. 4.13 Particle nucleating at a grain edge is bounded by three spherical surfaces: (a) General view, and (b) section normal to the grain edges.

particle nucleating on a grain corner is bounded by four spherical caps, Fig. 4.14. The free energy during nucleation ΔG^*_{het} at a grain edge or a grain corner is the same as in Eq. 4.44, except that we use V^* as the critical volume of the particle bounded by three or four spherical surfaces, respectively. In Fig. 4.12, the ratio $\Delta G^*_{\text{het}}/\Delta G^*_{\text{homo}}$ is plotted as a function of cos θ

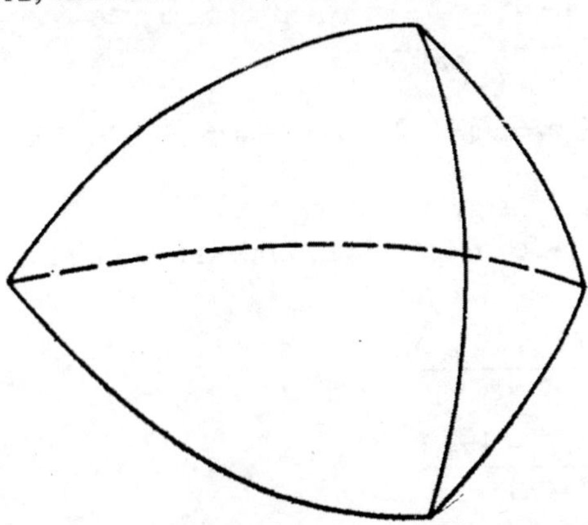

Fig. 4.14 Particle nucleating at a grain corner is bounded by four spherical surfaces.

for nucleation on grain edges and grain corners (alongwith that for grain boundary nucleation).

Nucleation on Inclusions

In the case of nucleation on the planar surface of a foreign particle (inclusion), the shape of the nucleating particle is a spherical cap or a half-lens. Here,

$$\Delta G^*_{\text{het}} = \tfrac{1}{4}\Delta G^*_{\text{homo}} (2 - 3 \cos \theta + \cos^3 \theta) \tag{4.49}$$

Note that the free energy of nucleation here is one-half of that of the full-lens case given in Eq. 4.43.

The following special cases may be noted:

1 $\theta = 180°$, $\cos \theta = -1$, $\Delta G^*_{\text{het}} = \Delta G^*_{\text{homo}}$. The product particle makes only point contact with the inclusion, which plays no role in the nucleation process.

2 $\theta = 0°$, $\cos \theta = 1$, $\Delta G^*_{\text{het}} = 0$. The product particle completely wets the inclusion surface and spreads out as a vanishingly thin film. Here, there is no barrier to nucleation.

3 $\theta = 90°$, $\cos \theta = 0$, $\Delta G^*_{\text{het}} = \tfrac{1}{2}\Delta G^*_{\text{homo}}$. The product particle is hemispherical in shape.

Nucleation on Dislocations

Cahn has described the nucleation process on dislocations based on the following assumptions:

1 The product particle β is cylindrical in shape extending along a straight dislocation line.

2 The strain energy of the dislocation assisting the nucleation process is taken to be removed over a distance from the dislocation equal to the radius r of the cylindrical particle.

3 An incoherent interface is formed.

4 The strain energy due to the particle formation can be neglected.

Then, per unit length of the dislocation, we have

$$\Delta G_{\text{het}} = \pi r^2 \Delta g + 2\pi r\sigma - A' \ln r \tag{4.50}$$

where $A' \ln r$ approximates the strain energy of the dislocation per unit length. $A' = \mu b^2/(4\pi(1 - \nu))$ for an edge dislocation and $\mu b^2/(4\pi)$ for a screw dislocation, where b is the Burgers vector of the dislocation and ν is Poisson's ratio of the matrix.

Setting $d(\Delta G)/dr = 0$, we obtain

$$2\pi r \Delta g + 2\pi\sigma - A'/r = 0$$

which is a quadratic equation in r for the critical condition. The two roots are:

$$r^*_{1,2} = -\frac{\sigma}{2\Delta g}\left(1 \pm \sqrt{\frac{1 + 2A'\Delta g}{\pi\sigma^2}}\right) \tag{4.51}$$

When $(1 + 2A'\Delta g)/\pi\sigma^2$ is positive, two real roots exist. From the sign of the second derivative, $d^2(\Delta G)/dr^2$, we note

$$r_1^*(\text{for } \Delta G_{\min}) = -\frac{\sigma}{2\Delta g}\left(1 - \sqrt{\frac{1 + 2A'\Delta g}{\pi\sigma^2}}\right) \tag{4.52}$$

$$r_2^*(\text{for } \Delta G_{\max}) = -\frac{\sigma}{2\Delta g}\left(1 + \sqrt{\frac{1 + 2A'\Delta g}{\pi\sigma^2}}\right) \tag{4.53}$$

So, a metastable minimum exists at $r = r_1^*$. Thereafter, a barrier maximum occurs at $r = r_2^*(r_2^* > r_1^*)$ in the ΔG versus r curve, Fig. 4.15.

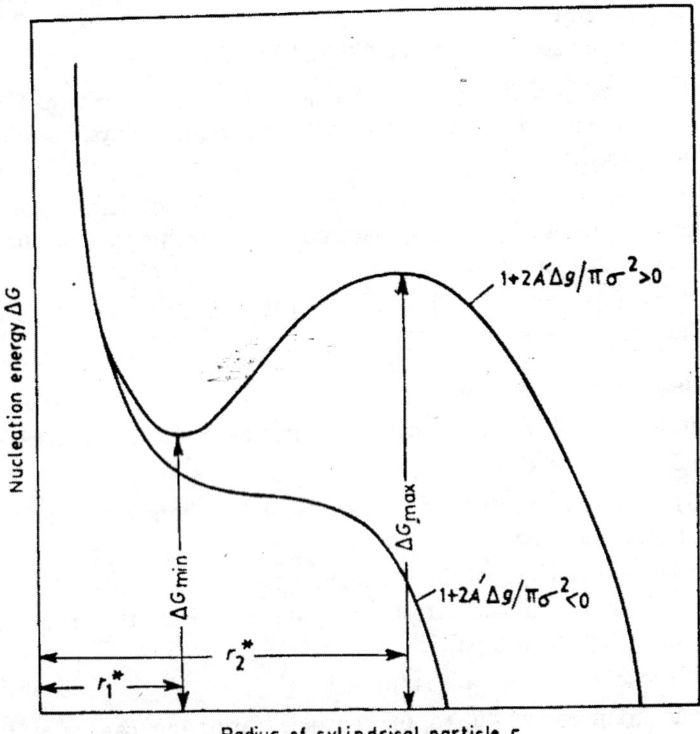

Fig. 4.15 The ΔG versus r curve for a cylindrical particle of radius r nucleating on a dislocation line.

Since Δg is negative, $(1 + 2A'\Delta g)/\pi\sigma^2$ can also become negative, especially at large driving forces. The roots $r_{1,2}^*$ are then imaginary. Neither a minimum nor a maximum exists in the ΔG versus r curve, Fig. 4.15. Under such conditions, there is no barrier to nucleation.

Interaction of Nucleation with Vacancies

A supersaturation or an undersaturation of vacancies can assist (or inhibit) the nucleation process under certain circumstances. If nucleation occurs incoherently, vacancies can be created or annihilated at the interface. If the incoherent product particle has a larger specific volume than the matrix, a supersaturation of vacancies in the matrix can assist in relieving the strain energy due to volume increase. Large supersaturations of vacancies are possible by quenching from high temperatures. The extra driving force contributed by the vacancy supersaturation may be of the order of the chemical driving force.

When a volume decrease occurs during the transformation, an undersaturation of vacancies can assist the nucleation process in a similar manner.

Rate of Heterogeneous Nucleation

The rate of heterogeneous nucleation can be expressed in a form similar to that of Eq. 4.9. In addition to the difference in the ΔG^* term as discussed above, the pre-exponential term will include only the number of preferred nucleation sites, which is many orders of magnitude smaller than the number of atoms per unit volume used in the homogeneous case. In order of magnitude, the number of nucleation sites for various situations is typically as follows:

$$N_{\text{homogeneous}} \sim 10^{29}\ \text{m}^{-3}$$

$$N_{\text{grain boundary}} \sim 10^{23}\ \text{m}^{-3}$$

$$N_{\text{grain edge}} \sim 10^{17}\ \text{m}^{-3}$$

$$N_{\text{grain corner}} \sim 10^{11}\ \text{m}^{-3}$$

$$N_{\text{dislocation}} \sim 10^{13}\ \text{m}^{-3}$$

In addition, we may have a distribution of sites of varying potencies (the greater the potency, the lower is the corresponding ΔG^*). If N_i is the number of sites per unit volume having a nucleation barrier ΔG_i^*, then from Eq. 4.9, we can write

$$I_{\text{het}} = \Sigma N_i \nu s_i^* \exp\left[-\frac{(\Delta G_i^* + \Delta G_D)}{RT}\right] \tag{4.54}$$

Usually, $I_{\text{het}} > I_{\text{homo}}$, because $\Delta G^*_{\text{het}} < \Delta G^*_{\text{homo}}$ dominates the result through the exponential factor, even though the pre-exponential factor may be much smaller for heterogeneous nucleation as compared to the homogeneous case.

FURTHER READING

J.W. Christian, *The Theory of Transformations in Metals and Alloys*, The Classical Theory of Nucleation, Chapter 10, p. 418 (1975).

F.R.N. Nabarro, *Proc. Phys. Soc.*, **52**, 90 (1940).

J.C. Fisher, J.H. Hollomon and D. Turnbull, *Trans. Amer. Min. Met. Engrs.*, **185**, 691 (1949).

J.D. Eshelby, *Proc. Roy. Soc.*, **241A**, 376 (1957).

P.J. Clemm and J.C. Fisher, *Acta Metall.*, **3**, 70 (1955).

J.W. Cahn, *Acta Metall.*, **5**, 169 (1957).

EXERCISES

4.1 Show that the homogeneous nucleation barrier ΔG^* for a spherical particle of radius r of a new phase is given by

$$\Delta G^* = \frac{16\pi\sigma^3}{3(\Delta g)^2}$$

neglecting strain energy effects. Also, show that the critical radius r^* of the nucleus is given by

$$r^* = -\frac{2\sigma}{\Delta g}.$$

4.2 Show that the critical values of the parameters for the nucleation of a thin disc under dilatational strains are

$$c^* = -\frac{2\sigma}{\Delta g}$$

$$r^* = 2\pi\mu \left(\frac{\Delta V}{V}\right)^2 \frac{\sigma}{(\Delta g)^2}$$

$$\Delta G^* = \frac{8\pi^3\mu^2}{3}\left(\frac{\Delta V}{V}\right)^4 \frac{\sigma^3}{(\Delta g)^4}$$

Assume an incoherent interface between the parent and the product phases.

4.3 In a given nucleation process in an iron-base alloy, $\Delta g = -210$ MJ m^{-3}, the interfacial energy is 0.02 J m^{-2} and the fractional change in volume is 0.04. Calculate the dimensions of the critical nucleus and the number of atoms in the nucleus. Would you expect particles of this size to be detectable in the electron microscope?

4.4 Show that the nucleation barrier ΔG^* for the case of a shear transformation, when the product phase is oblate-spheroidal shaped, is given by

$$\Delta G^* = \frac{32\pi}{3} \frac{A^2\sigma^3}{(\Delta g)^4}.$$

4.5 In cobalt, a coherent interface forms during HCP to FCC transformation. The lattice parameter of the FCC phase is 3.536 Å. The distance between nearest neighbours along a close packed direction in the basal plane of the HCP phase is 2.507 Å. Show that the misfit to be accommodated by coherency strains is small.

4.6 Show that (a) the volume of a spherical cap is

$$V = \frac{\pi}{3} r^3(2 - 3\cos\theta + \cos^3\theta)$$

(b) The surface area S_A of a spherical cap is

$$S_A = 2\pi r^2(1 - \cos\theta).$$

4.7 If the surface energies $\sigma_{\alpha\beta}$, $\sigma_{\beta\gamma}$, and $\sigma_{\alpha\alpha}$ are equal to one another, show that the work needed to form a nucleus of β on a plane surface of γ is $\frac{1}{2}\Delta G^*_{\text{homo}}$, if γ is not deformable and is $\frac{5}{16}\Delta G^*_{\text{homo}}$, if γ readily deforms.

4.8 When the activation energy for interfacial transfer of atoms is negligible, show that at $T_{I_{\max}}$

$$\left(\frac{4T_{I\max}}{\Delta g}\right)\left(\frac{d\Delta g}{dT}\right)_{T_{I\max}} + 1 = 0.$$

Assume that ΔG^* is as given in Exercise 4.4.

5

Growth Kinetics

In Chapter 4, the process of nucleation was identified with the addition of one atom to the critical sized nucleus to make it just supercritical. All *further* additions of atoms to the nucleus are considered to be steps in the process of growth. We shall consider here the kinetics of growth, under conditions where strain energy plays no part and heat flow does not control the growth process. Interface-controlled growth is considered first. The rate controlling step here is the transfer of atoms across the interface. Transformations involving no change in composition are interface-controlled. Some transformations with compositional changes can also be interface-controlled. In the later sections of this chapter, we consider diffusion-controlled growth. Here, the long-range diffusion of one of the component atoms through the matrix is rate controlling.

5.1 INTERFACE-CONTROLLED GROWTH

Growth Rate at Constant Temperature

We shall assume that the interface is non-singular and incoherent and that the atoms can cross it independently of one another at all points. Let ΔG_D be the activation energy for an atomic jump across the interface. The change in the chemical free energy per atom in the $\alpha \rightarrow \beta$ transformation is $v\Delta g$, where v is the volume per atom. The free energy barrier for an atomic jump from α to β (in the direction of the driving force) is ΔG_D, while the free energy barrier for an atomic jump from β back to α (against the driving force) is $\Delta G_D - v\Delta g$, as illustrated in Fig. 5.1. Note that $\Delta g < 0$ for a finite driving force. Thus the barrier in the reverse direction is larger than ΔG_D.

The net rate of atomic jumping from α to β per unit area of the interface is

$$\left(\frac{\mathrm{d}n^{\alpha\rightarrow\beta}}{\mathrm{d}t}\right)_{\text{net}} = \text{forward rate} - \text{reverse rate}$$

$$= s\nu \exp\left(-\frac{\Delta G_D}{kT}\right) - s\nu \exp\left[-\frac{(\Delta G_D - v\Delta g)}{kT}\right]$$

$$= s\nu \exp\left(-\frac{\Delta G_D}{kT}\right)\left[1 - \exp\left(\frac{v\Delta g}{kT}\right)\right] \tag{5.1}$$

where s is the number of interfacial atoms in each phase per unit area of the interface and ν is the lattice vibration frequency.

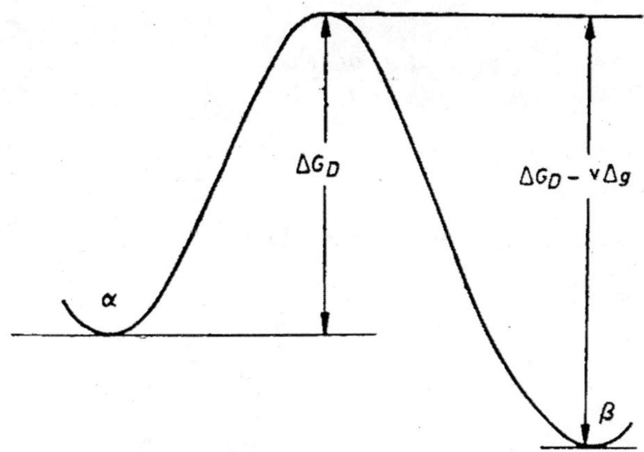

Fig. 5.1 The free energy barrier for atom jump across the interface during growth.

The growth rate of the β phase, $U(\mathrm{m\,s^{-1}})$, can then be expressed as

$$U = \frac{\mathrm{d}r}{\mathrm{d}t} = \frac{\mathrm{d}V}{\mathrm{d}t} \quad \text{(per m}^2\text{ of the interface)}$$

$$= \left(\frac{\mathrm{d}n^{\alpha\to\beta}}{\mathrm{d}t}\right)_{\text{net}} \cdot v$$

$$= sv\nu \exp\left(-\frac{\Delta G_D}{kT}\right)\left[1 - \exp\left(\frac{v\Delta g}{kT}\right)\right]$$

$$= \lambda_j \nu \exp\left(-\frac{\Delta G_D}{kT}\right)\left[1 - \exp\left(\frac{v\Delta g}{kT}\right)\right] \tag{5.2}$$

where λ_j is the jump distance across the interface. It is also the volume corresponding to 1 m² of the interface advancing through a single atomic growth step, i.e., $\lambda_j = sv$.

The diffusion coefficient D_b for the diffusional jump across the interface can be defined as

$$D_b = \lambda_j^2 \nu \exp\left(-\frac{\Delta G_D}{kT}\right) \tag{5.3}$$

D_b is usually taken to be the diffusivity in the boundary between the two phases.

Growth Rate as a Function of Temperature

The above expression for U shows that the growth rate is zero at T_0, where $\Delta g = 0$. Just below T_0, Δg is negative but very small. Thus, for a small driving force,

$$-v\Delta g \ll kT \tag{5.4}$$

$$\exp\left(\frac{v\Delta g}{kT}\right) \approx 1 + \frac{v\Delta g}{kT} \tag{5.5}$$

and

$$U \approx \lambda_j \nu \left(-\frac{v\Delta g}{kT}\right) \exp\left(-\frac{\Delta G_D}{kT}\right)$$

$$= \frac{D_b}{\lambda_j}\left(-\frac{v\Delta g}{kT}\right) \tag{5.6}$$

At temperatures where the driving force is large,

$$-v\Delta g \gg kT \tag{5.7}$$

$$\exp\left(\frac{v\Delta g}{kT}\right) \ll 1 \tag{5.8}$$

and

$$U \approx \lambda_j \nu \exp\left(-\frac{\Delta G_D}{kT}\right)$$

$$= \frac{D_b}{\lambda_j} \tag{5.9}$$

Thus the growth rate increases approximately linearly with Δg for very small degrees of supercooling. However, as the supercooling increases, the D_b term, which depends exponentially on temperature, decreases rapidly. The growth rate after reaching a maximum, decreases exponentially with temperature, as given by Eq. 5.9.

Experimental Measurements

The measurement of growth rates is more straightforward than that of nucleation rates. The size of the largest sphere (or nodule) of the growing particle is measured metallographically, using a series of progressively transformed specimens. These measurements can be carried out readily in samples, where the fraction transformed is 0.2 or less. The impingement of the growing particles on one another is usually negligible up to this stage of the transformation. The radius of the largest sphere seen on the plane of observation can be taken to be equal to the true radius of the three-dimensional particle. A plot of the maximum size against transformation time yields a straight line for interface-controlled transformations.

Figure 5.2 shows the temperature dependence of the linear growth rate for the transformation of white to gray tin ($T_0 = +13$°C). The maximum in the growth rate (~1 mm/hr) occurs at -32°C, where $\Delta T = 45$°C. Thereafter, the growth rate decreases exponentially with temperature as the boundary diffusivity D_b decreases, as given by Eq. 5.9.

5.2 DIFFUSION-CONTROLLED GROWTH

Diffusion-controlled growth refers to the growth process, where there is a change in composition and the rate is controlled by the long-range diffusion of one of the component atoms through the matrix phase.

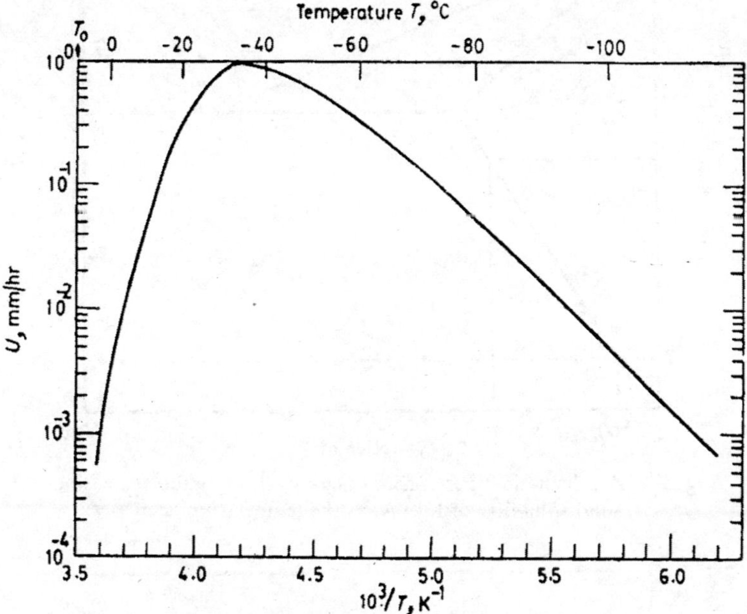

Fig. 5.2 The temperature dependence of the growth rate for the transformation of white to gray tin.

A General Expression for the Growth Rate

Consider the continuous precipitation of β phase particles from a supersaturated solid solution of α in a binary system of A and B. Let the compositions be expressed in terms of c, the concentration of B atoms in units of mol m^{-3}. Referring to the schematic phase diagram in Fig. 5.3, at temperature T_0, the α solid solution of composition $\bar{c}$ is just saturated. On cooling to temperature T, the solid solution becomes supersaturated with respect to the B component atoms. Precipitate particles of the β phase form in the matrix. The β particles being rich in B, the composition of the matrix changes towards the equilibrium value $c_{\alpha\beta}$ $(< \bar{c})$.

$$\underset{\bar{c}}{\alpha_{\text{supersat}}} \rightarrow \underset{c_{\alpha\beta}}{\alpha_{\text{sat}}} + \underset{c_{\beta\alpha}}{\beta\text{-precipitate}}$$

The concentration-distance profile during the precipitation is schematically illustrated in Fig. 5.4. Along the distance axis, at $y = 0$, the precipitation starts in the matrix of initial composition $\bar{c}$. The precipitate has grown to a size r, where the interface separating the precipitate from the matrix is located. The composition within the precipitate particle from $y = 0$ to $y = r$ is uniform and is equal to $c_{\beta\alpha}$. *Equilibrium is assumed to prevail at the interface*, so that the composition of the α matrix in contact with the precipitate

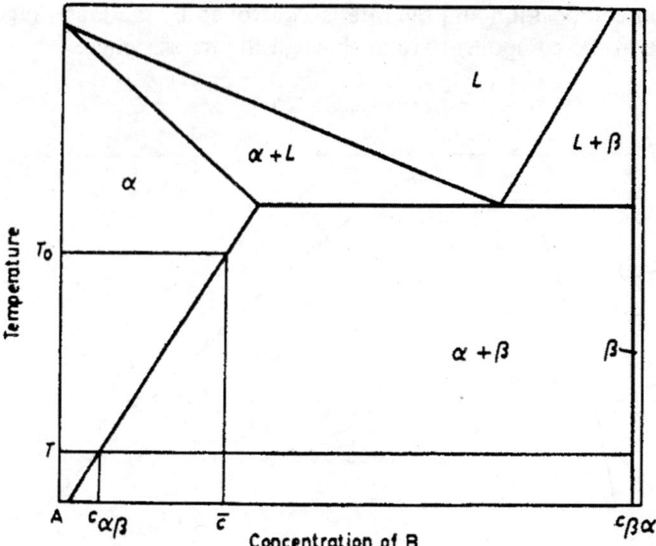

Fig. 5.3 A schematic binary phase diagram, showing the decreasing solid solubility of B in A, as the temperature decreases.

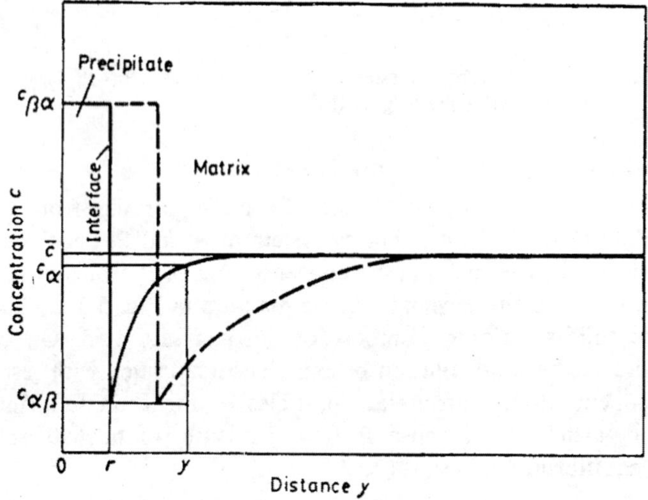

Fig. 5.4 The concentration of B atoms as a function of distance during continuous precipitation of β.

particle is $c_{\alpha\beta}$. Due to the formation of the B-rich precipitate, there is a depletion of B atoms in the matrix immediately surrounding it. The matrix composition becomes the initial value of $\bar{c}$ only at large distances. The B atoms diffuse down the concentration gradient set up by the depletion, arrive at the interface and cross it into the precipitate. As the precipitate particle grows to larger sizes, the region of the surrounding matrix depleted of B atoms extends to larger distances, as shown by the dotted-line profile in

Fig. 5.4. c_α denotes the concentration of B atoms in the depleted region. It is a function of the diffusion time t and distance y.

As the composition at the interface remains independent of time, the number of B atoms that arrive from the matrix at the interface (per m^2 of the interface) must be equal to the number of B atoms that are added to the growing particle (per m^2 of the interface). Based on this conservation condition, we can write

$$U(c_{\beta\alpha} - c_{\alpha\beta}) = D_\alpha \left(\frac{\partial c_\alpha}{\partial y}\right)_{y=r} \tag{5.10}$$

where $U(= \mathrm{d}r/\mathrm{d}t)$ is the growth rate of the particle, D_α is the diffusion coefficient of B in the α matrix, and $D_\alpha(\partial c_\alpha/\partial y)_{y=r}$ is the flux of B atoms at the interface due to the concentration gradient in the matrix phase. Hence, we can write a general expression for the growth rate as

$$U = \frac{D_\alpha\left(\frac{\partial c}{\partial y}\right)_{y=r}}{c_{\beta\alpha} - c_{\alpha\beta}}. \tag{5.11}$$

Growth of Spherical Particles

For spherical particles, the growth distance r is the radius of the growing sphere. It is assumed that the solubility of B atoms in the α matrix at the interface is independent of r. (At small values of r, the solubility *is* a function of r and this case is discussed separately a little later.)

The concentration profile in the matrix can be approximated to the following equation:

$$c_\alpha = \bar{c} - (r/y)(\bar{c} - c_{\alpha\beta}) \tag{5.12}$$

This yields

$$c_\alpha = c_{\alpha\beta} \quad \text{when } y = r$$

and $$c_\alpha = \bar{c} \quad \text{when } y = \infty$$

At distances $\infty > y > r$, c_α varies between $\bar{c}$ and $c_{\alpha\beta}$. Also, by differentiating Eq. 5.12 with respect to y; we obtain:

$$\left(\frac{\partial c}{\partial y}\right)_{y=r} = \left(\frac{\bar{c} - c_{\alpha\beta}}{r}\right) \tag{5.13}$$

Substituting Eq. 5.13 in 5.11, we get

$$U = \frac{D_\alpha(\bar{c} - c_{\alpha\beta})}{r(c_{\beta\alpha} - c_{\alpha\beta})} \tag{5.14}$$

Substituting $U = \mathrm{d}r/\mathrm{d}t$ and rearranging, we get

$$r\,\mathrm{d}r = \frac{D_\alpha(\bar{c} - c_{\alpha\beta})}{(c_{\beta\alpha} - c_{\alpha\beta})}\,\mathrm{d}t \tag{5.15}$$

As we have assumed here that $c_{\alpha\beta}$ is not a function of r, integration of Eq. 5.15 yields

$$r^2 = \frac{2D_\alpha(\bar{c} - c_{\alpha\beta})t}{(c_{\beta\alpha} - c_{\alpha\beta})} \tag{5.16}$$

$$r \propto t^{1/2}$$

$$U \propto t^{-1/2} \tag{5.17}$$

The radius of the particle increases as the square root of time, signifying *parabolic growth.* As there is a progressive depletion of B atoms in the matrix, the growth *rate U* decreases with time.

Growth of Small Spherical Particles

When r is very small, $c_{\alpha\beta}$ does depend on r. The Thomson-Freundlich equation describes this dependence:

$$c^r_{\alpha\beta} = c^\infty_{\alpha\beta} \exp\left(\frac{2V\sigma}{RTr}\right) \tag{5.18}$$

where $c^r_{\alpha\beta}$ and $c^\infty_{\alpha\beta}$ are the solute concentrations in the α matrix in equilibrium with a β particle of radius r and ∞, respectively.

We can recalculate the growth kinetics when Eq. 5.18 is applicable. If $2V\sigma/RTr \ll 1$, we can approximate Eq. 5.18 to

$$c^r_{\alpha\beta} \approx c^\infty_{\alpha\beta}\left(1 + \frac{2V\sigma}{RTr}\right) \tag{5.19}$$

If we take that $c_{\alpha\beta} \to \bar{c}$ when $r \to r^*$, where r^* is the critical size for nucleation, we have

$$\frac{\bar{c} - c^r_{\alpha\beta}}{\bar{c} - c^\infty_{\alpha\beta}} = \frac{c(\infty)\left(1 + \dfrac{2V\sigma}{RTr^*}\right) - c(\infty)\left(1 + \dfrac{2V\sigma}{RTr}\right)}{c(\infty)\left(1 + \dfrac{2V\sigma}{RTr^*}\right) - c(\infty)}$$

$$= r^*\left(\frac{1}{r^*} - \frac{1}{r}\right)$$

$$= \left(1 - \frac{r^*}{r}\right) \tag{5.20}$$

As $c_{\alpha\beta}$ is usually small as compared with $c_{\beta\alpha}$, we can take that $c_{\beta\alpha} - c^r_{\alpha\beta} \approx c_{\beta\alpha} - c^\infty_{\alpha\beta}$. The equation for the growth rate U then becomes

$$U = \frac{D(\bar{c} - c^\infty_{\alpha\beta})}{r(c_{\beta\alpha} - c_{\alpha\beta})}\left(1 - \frac{r^*}{r}\right) \tag{5.21}$$

1 U is negative, when $r < r^*$; such particles are subcritical and tend to dissolve

2 $U = 0$ for $r = r^*$; such particles are in unstable equilibrium with the matrix of composition $\bar{c}$

3 U is positive for $r > r^*$; such particles are stable and tend to grow.

The maximum growth rate, U_{max}, is obtained by setting $dU/dr = 0$ and occurs at $r = 2r^*$:

$$U_{max(r=2r^*)} = \frac{D}{4r^*} \frac{\bar{c} - c_{\alpha\beta}^{\infty}}{c_{\beta\alpha} - c_{\alpha\beta}} \tag{5.22}$$

The variation of U with r is schematically shown in Fig. 5.5. At $r \gg r^*$, the parabolic growth rate, derived for the case where $c_{\alpha\beta}$ is not a function of r, prevails.

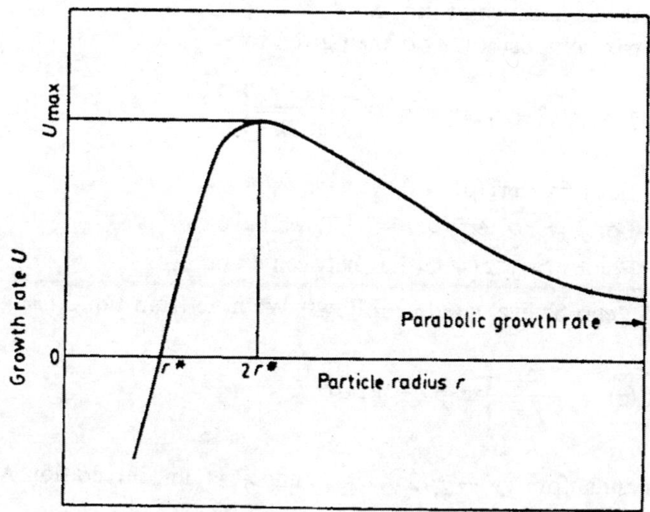

Fig. 5.5 Schematic variation of the growth rate of a particle of radius r, r being in the range of r^*.

The temperature dependence of U_{max} can also be estimated. If we assume that the solvus is a straight line as in Fig. 5.3, i.e., the solubility decreases linearly with temperature, we have

$$\bar{c} - c_{\alpha\beta}^{\infty} \propto (T_0 - T) \tag{5.23}$$

where T is the reaction temperature. From the nucleation equations, we know that

$$r^* \propto \frac{1}{(T_0 - T)} \tag{5.24}$$

Using these in conjunction with the exponential dependence on temperature of D, we obtain from Eq. 5.22,

$$U_{max} \propto (T_0 - T)^2 \exp\left(-\frac{Q}{RT}\right) \tag{5.25}$$

Thus, $U_{max} = 0$ at $T = T_0$ and is also very small at low temperatures.

The temperature dependence of the maximum growth rate can be obtained by setting $dU_{r=2r^*}/dT = 0$. This yields the condition

$$2T^2 + \frac{QT}{R} - \frac{QT_0}{R} = 0 \tag{5.26}$$

Plane Front Growth

Here, the interface between the precipitate and the matrix is planar. The radius of curvature of the interface is infinite. So, the concentration of B atoms in the matrix in contact with the precipitate is $c_{\alpha\beta}^{\infty}$ and is independent of the growth distance r and time t.

The matrix composition c_α is then given by

$$c_\alpha = \bar{c} - (\bar{c} - c_{\alpha\beta}^{\infty})\left[1 - \text{erf}\,\frac{y-r}{2\sqrt{D_\alpha t}}\right] \tag{5.27}$$

1 For $y = r$, erf (0) = 0, we have $c_\alpha = c_{\alpha\beta}^{\infty}$

2 For $y = \infty$, erf $(\infty) = +1$, we have $c_\alpha = \bar{c}$

3 For $\infty > y > r$, c_α lies between $\bar{c}$ and $c_{\alpha\beta}^{\infty}$

$(\partial c/\partial y)_{y=r}$ can be evaluated as follows. We note from Eq. 2.13,

$$\text{erf}\,(z) = \frac{2}{\sqrt{\pi}}\int_0^z \exp\,(-\eta^2)\,d\eta$$

where z stands for $(y-r)/2\sqrt{D_\alpha t}$ and η is an integration variable. Differentiating erf (z) with respect to z,

$$\frac{d\,\text{erf}\,(z)}{dz} = \frac{d}{dz}\left\{\frac{2}{\sqrt{\pi}}[f(z) - f(0)]\right\}$$

$$= \frac{2}{\sqrt{\pi}}\exp\,(-z^2) \tag{5.28}$$

where $f(z)$ is some function of z.

$$\frac{d\,\text{erf}\,(z)}{dy} = \frac{d\,\text{erf}\,(z)}{dz}\,\frac{dz}{dy} \tag{5.29}$$

Substituting for z and for d erf (z)/dz from Eq. 5.28 into 5.29, we obtain

$$\frac{d\,\text{erf}\left(\dfrac{y-r}{2\sqrt{D_\alpha t}}\right)}{dy} = \frac{1}{\sqrt{\pi D_\alpha t}}\exp\left[-\frac{(y-r)^2}{2\sqrt{D_\alpha t}}\right] \tag{5.30}$$

So, differentiating Eq. 5.27 with respect to r, we get

$$\left(\frac{\partial c}{\partial y}\right)_{y=r} = \frac{\bar{c} - c_{\alpha\beta}}{\sqrt{\pi D_\alpha t}} \tag{5.31}$$

Substituting Eq. 5.31 in the general growth equation (Eq. 5.11), we have

$$U = \frac{dr}{dt} = \sqrt{\frac{D_\alpha}{\pi t}}\left(\frac{\bar{c} - c_{\alpha\beta}}{c_{\beta\alpha} - c_{\alpha\beta}}\right) \tag{5.32}$$

Integrating Eq. 5.32,

$$r = 2\sqrt{\frac{D_\alpha}{\pi}}\left(\frac{\bar{c} - c_{\alpha\beta}}{c_{\beta\alpha} - c_{\alpha\beta}}\right)t^{1/2} \tag{5.33}$$

This indicates parabolic growth rate, as in the case of growth of spherical particles with $r \gg r^*$.

FURTHER READING

J.W. Christian, *The Theory of Transformations in Metals and Alloys*, Chapter 11, Theory of Thermally Activated Growth, Pergamon Press, Oxford, p. 476 (1975).

EXERCISE

5.1 For a diffusion-controlled precipitation process, $T_0 = 725°C$. The activation energy for diffusion is 100 kJ mol^{-1}. Compute the temperature at which the growth rate of nuclei with $r = 2r^*$ is a maximum.

6

Overall Transformation Kinetics

The overall transformation kinetics describe the relationship between fraction transformed and time for different reaction temperatures. It is a function of both the nucleation and the growth rates. Nucleation is necessary, before growth can occur. Growth of the nucleated particles at an appreciable rate is necessary for obtaining a significant fraction of the transformed volume.

The final grain size of the product depends on the nucleation and the growth rates. A high nucleation rate coupled with a low growth rate yields a large number of crystals and a fine grain size. On the other hand, a low nucleation rate and a high growth rate yield fewer crystals and a coarse grain size.

In this chapter, the empirical equations describing the transformation kinetics are first introduced. The Johnson-Mehl-Avrami models are then discussed, followed by the transformation kinetics for special nucleation sites. The process of continuous precipitation is used as an example of a transformation characterized by diffusion-controlled growth.

6.1 EMPIRICAL EQUATIONS

The empirical equations describing the transformation kinetics are either of the exponential form or the power form. In the *exponential form*, the volume fraction transformed X as a function of time t is expressed as:

$$X = 1 - \exp[-(Kt)^n] \tag{6.1}$$

where K is the *reaction rate constant* (in units of s^{-1}) and n is the *time exponent* (dimensionless). Taking logarithms in Eq. 6.1, we have

$$\log \frac{1}{1 - X} = (Kt)^n \tag{6.2}$$

Taking logarithms again, we have

$$\log \log \frac{1}{1 - X} = n \log K + n \log t \tag{6.3}$$

Experimental data of fraction transformed versus time can be plotted as log log $1/(1 - X)$ against log t. A straight line is obtained, if Eq. 6.1 is valid. The slope of the straight line yields the time exponent n. K can be obtained from the intercept on the y-axis. Such a plot is shown for the $\beta \rightarrow \alpha$ manganese transformation in Fig. 6.1 at four transformation temperatures. Values of 0 and -2 on the y-axis correspond to $X = 0.9$ and 0.02, respectively.

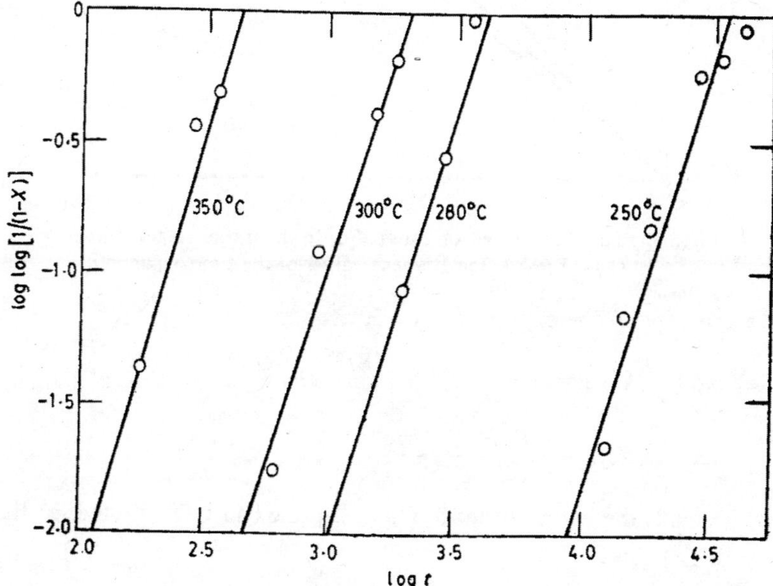

Fig. 6.1 Log log $1/(1 - X)$ versus log t plot for $\beta \rightarrow \alpha$ transformation in manganese (after Christian).

Differentiating Eq. 6.1 with respect to time, we obtain the transformation rate:

$$\frac{dX}{dt} = \exp[-(Kt)^n] nK^n t^{n-1}$$

Substituting for $(1 - X)$ from Eq. 6.1, we have

$$\frac{dX}{dt} = nK^n t^{n-1}(1 - X) \tag{6.4}$$

The nature of X versus t curves depends on the value of n. If $n = 1$, $dX/dt = K(1 - X)$ and the curve becomes a straight line at small values of X with the slope of K, Fig. 6.2. For $n > 1$, the transformation rate increases with time, see Fig. 6.2. For $1 > n > 0$, the transformation rate decreases with time. $n < 0$ is a situation not realizable physically.

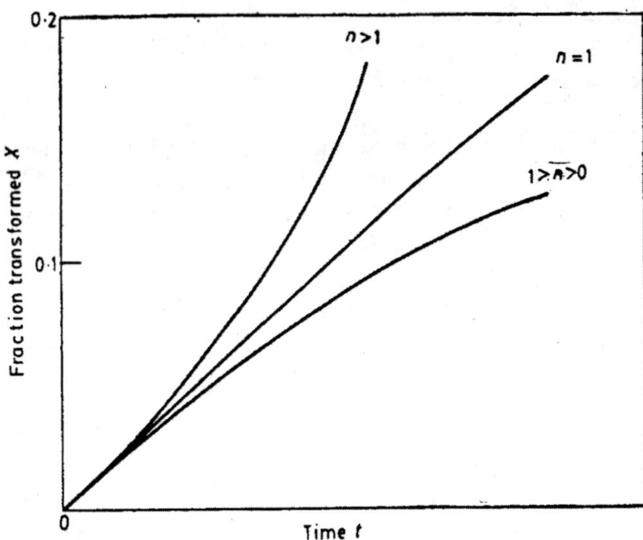

Fig. 6.2 The transformation rate dX/dt in the initial stages when $(1 - X) \sim 1$, for different values of the time exponent n.

We can expand the exponential of Eq. 6.1 as

$$\exp[-(Kt)^n] = 1 - (Kt)^n + \frac{(Kt)^{2n}}{2!} - \frac{(Kt)^{3n}}{3!} + \ldots \tag{6.5}$$

or

$$X = (Kt)^n - \frac{(Kt)^{2n}}{2!} + \frac{(Kt)^{3n}}{3!} - \ldots \tag{6.6}$$

If (Kt) is small, the higher order terms in Eq. 6.6 can be neglected, so that

$$X = (Kt)^n \tag{6.7}$$

This is known as the *power form* describing the transformation kinetics. The corresponding transformation rate is

$$\frac{\mathrm{d}X}{\mathrm{d}t} = nK^n t^{n-1} \tag{6.8}$$

Comparison of Eqs. 6.4 and 6.8 shows that the only difference between these two empirical rate laws is the factor $(1 - X)$, the latter being the fraction of the untransformed phase. The factor $(1 - X)$ is called the *impingement factor*, because it takes into account the interference among the growing particles of the transformation product. Accordingly, the exponential form is considered to be corrected for impingement, while the simplified power form is not.

6.2 TRANSFORMATION KINETICS FOR INTERFACE-CONTROLLED GROWTH

The Johnson-Mehl Model

The Johnson-Mehl model of the transformation kinetics for nucleation and growth processes is based on the following assumptions:

1 The nucleation rate I is equal to $1/(1 - X)\, \mathrm{d}N_v/\mathrm{d}t$, when N_v is the number of nucleating particles per unit volume of the system and the factor $1/(1 - X)$ relates the nucleation rate to a unit volume of the untransformed phase. I is taken to be independent of time t. This assumption ignores any transient effects associated with the attainment of the equilibrium distribution of embryos (refer to Chapter 4, p. 51. It also ignores any increase in I due to autocatalysis.

2 The growth rate $U(= \mathrm{d}r/\mathrm{d}t)$ is also taken to be independent of X and t. This assumption is valid for interface-controlled growth (refer to Chapter 5, p. 72).

3 Nucleation occurs randomly in the untransformed phase. This assumption does not necessarily mean homogeneous nucleation. Preferred nucleation sites may be randomly distributed in the parent phase. Nonrandom distribution of preferred sites such as segregated foreign particles or inclusions or the close proximity of potential nucleation sites on a grain boundary are not provided for in the model. Such a nonrandom distribution can lead to an effect called *site saturation* discussed later in connection with special nucleation sites on p. 87.

4 Product phase particles grow as spheres until impingement occurs. The assumption of spherical growth is valid, provided the growth rate is independent of crystal direction, i.e.,

$$U_x = U_y = U_z.$$

At time t, the volume of a spherical particle that nucleated at time $\tau (\tau < t)$ is equal to $\frac{4}{3}\pi U^3(t - \tau)^3$. The number of particles nucleated per unit volume of the alloy in time $\mathrm{d}\tau$ is $I\,\mathrm{d}\tau$. So, the volume of all these particles at time t is equal to the corresponding volume fraction transformed:

$$\mathrm{d}X_{\mathrm{ext}} = \frac{4}{3}\pi U^3(t - \tau)^3 I\, \mathrm{d}\tau \tag{6.9}$$

The subscript 'ext' stands for 'extended' and denotes that the transformed volume is not corrected for impingement of the growing spheres. Introducing the impingement factor $(1 - X)$, we have

$$\mathrm{d}X_{\mathrm{ext}}(1 - X) = \mathrm{d}X_{(\text{corrected for impingement})} \tag{6.10}$$

Combining Eqs. 6.9 and 6.10, we have for the corrected fraction:

$$\frac{\mathrm{d}X}{(1 - X)} = \frac{4}{3}\pi U^3(t - \tau)^3 I\, \mathrm{d}\tau \tag{6.11}$$

Integrating Eq. 6.11, we obtain the Johnson-Mehl equation:

$$\int_0^X \frac{\mathrm{d}X}{(1 - X)} = \int_{\tau=0}^{\tau=t} \frac{4}{3}\pi U^3(t - \tau)^3 I\, \mathrm{d}\tau$$

or

$$X = 1 - \exp\left(-\frac{\pi}{3} IU^3t^4\right) \tag{6.12}$$

Here, the time exponent n is 4 and the reaction rate constant K is $\left(\frac{\pi}{3} IU^3\right)^{1/4}$. The value of $n = 4$ indicates that the volume of the transformation product increases (1) with the first power of time, due to the constant rate of nucleation, and (2) with the third power of time due to constant *linear* growth rate, dr/dt, as volume is proportional to r^3. A plot of X versus t for typical values of $I(= 10^9\ m^{-3}\ s^{-1})$ and $U(= 3 \times 10^{-7}\ m\ s^{-1})$ shown in Fig. 6.3 has the sigmoidal shape.

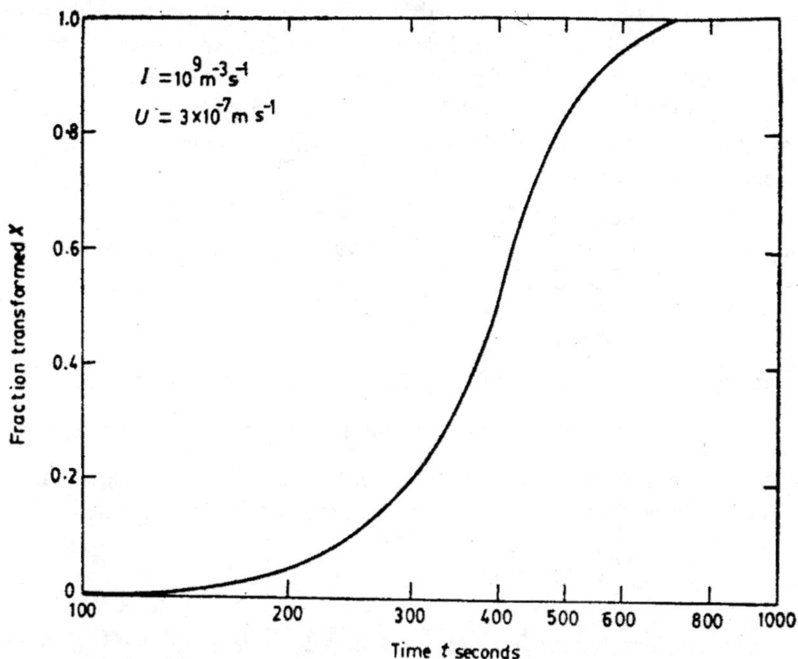

Fig. 6.3 Fraction transformed X as a function of reaction time t for typical values of nucleation rate I and growth rate U.

The Avrami Model

Avrami assumed that there are N_v preferred sites per unit volume and they decrease exponentially with time on nucleation:

$$N_t = N_v \exp(-\omega t) \tag{6.13}$$

where N_t is the number of preferred nucleation sites left over at time t and ω is a frequency factor characteristic of a particular system. If ω is very large, even for small t, $\exp(-\omega t) \sim 0$, so that all sites are nucleated at the start of the transformation process. Under such conditions, the fraction transformed X varies with time t as follows:

$$X = 1 - \exp\left[-\frac{\pi}{3}N_v U^3 t^3\right] \tag{6.14}$$

Note that the time exponent here turns out to be three, arising from a constant linear growth rate, the nucleation rate I being zero.

If ω is small, a decreasing rate of nucleation is observed during the transformation. Correspondingly, the time exponent under such conditions varies between 4 and 3.

Transformation Kinetics for Special Nucleation Sites

The Johnson-Mehl-Avrami equations may not always be applicable, if nucleation occurs at special sites such as grain boundaries, grain edges or grain corners. Consider the nucleation at such sites for the case of spherical growth at a constant linear growth rate. Grain corners, for example, can be considered to be randomly distributed within the material. If the nucleation rate is relatively small and constant, Eq. 6.12 may be applicable. If the nucleation rate is large, all the grain corners may be exhausted in the early stages of the transformation. I can be taken to be zero during the transformation and Eq. 6.14 due to Avrami may apply. The term N_v in Eq. 6.14 will then be equal to the number of grain corners per unit volume of the parent phase.

In the case of grain boundary and grain edge nucleation, the preferred sites cannot be regarded as randomly distributed. For example, two particles nucleating along the same grain edge have a much higher probability of impinging on each other than with particles growing out from another edge. If the sites are exhausted at an early stage, a rod-like particle tends to grow along each grain edge, as schematically shown in Fig. 6.4(a). Similarly, in the case of grain boundary nucleation, a high nucleation rate can result in early site saturation. The product then grows in a slab-like shape, Fig. 6.4(b).

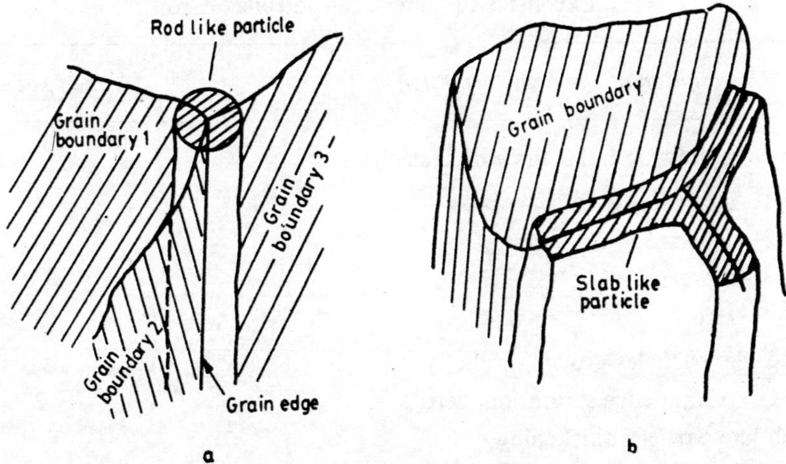

Fig. 6.4 (a) Growth of a rod-like particle from a grain edge, and (b) growth of a slab-like particle from a grain boundary.

These geometries lead to the following kinetic equations, which are significantly different from the Johnson-Mehl-Avrami equations. The rod-like particle from a grain edge grows radially in two dimensions, giving rise to a time exponent of 2 with a constant linear growth rate U and with $I \sim 0$ during the transformation:

$$X = 1 - \exp(-\pi L_v U^2 t^2) \tag{6.15}$$

where L_v is the total length of grain edges in unit volume.

The slab-like particle at a grain boundary grows in the thickness direction, leading to a time exponent of 1 with constant U and $I \sim 0$:

$$X = 1 - \exp(-2S_v Ut) \tag{6.16}$$

where S_v is the grain boundary area per unit volume. The effect of grain size is also described by a simple kinetic law:

$$t_f \approx \frac{1}{2} \frac{\bar{D}}{U} \tag{6.17}$$

where t_f is the time for virtual completion of the transformation and $\bar{D}$ is the average grain diameter.

Time Exponents in Interface-controlled Growth

For interface-controlled growth, a time exponent of 1 arises for each *growing* linear dimension of the particle. For a constant rate of nucleation, an additional exponent of 1 is added. For zero rate of nucleation, there is no addition to the time exponent. Table 6.1 lists the time exponents for different situations under which interface-controlled growth occurs.

TABLE 6.1
Time Exponents for Interface-controlled Growth

Description of the transformation	*Time exponent*
Particles growing in all three dimensions	
1 increasing I	> 4
2 constant I	4
3 decreasing I	3-4
4 zero I	3
Rod-like particles growing radially	
Grain-edge saturation, zero I	2
Slab-like particles thickening	
Grain boundary saturation, zero I	1

6.3 TRANSFORMATION KINETICS FOR DIFFUSION-CONTROLLED GROWTH

Kinetics of Continuous Precipitation

Figure 6.5 shows the amount of precipitation as a function of time at 400°C in a Au-30 wt. % Ni alloy for three different grain sizes, expressed as grain boundary area per unit volume (S_v). The rate of precipitation is faster for finer grain sizes, as the precipitates are nucleated at the grain boundaries of the parent phase.

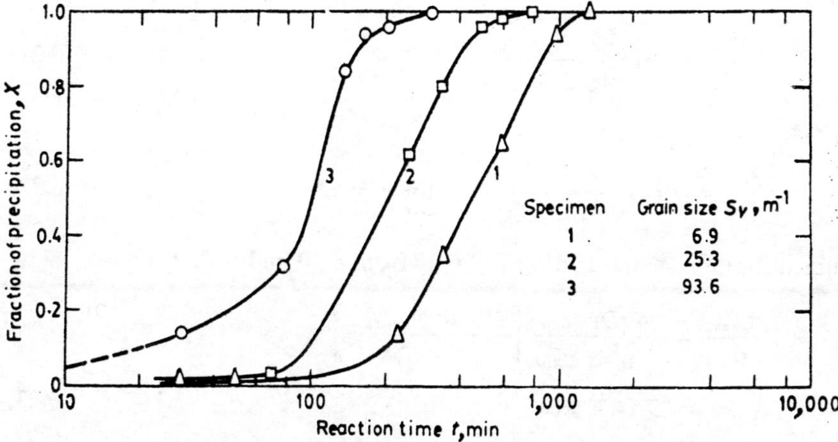

Fig. 6.5 Fraction of precipitate X as a function of time at 400°C during continuous precipitation in a Au-30 wt.% Ni alloy for three different grain sizes.

We can derive expressions for the precipitation kinetics, X versus t, for the case of spherical growth with a compositional change, where diffusion in the matrix is rate controlling. Let the transformation be the continuous precipitation of β particles from a supersaturated α matrix:

$$\alpha_{\text{supersat}} \rightarrow \alpha_{\text{sat}} + \beta$$

Let us assume that all the nucleation centres per unit volume N_v come into existence at $t \sim 0$ so that $I = 0$ during the transformation. The transformation is complete as soon as all the excess solute has precipitated as β particles and the depleted α matrix attains the uniform equilibrium composition $c_{\alpha\beta}$. For this situation, the fraction transformed X is defined as the number of moles of B atoms in the transformation product at any time t, divided by the total number of moles of B available for the transformation:

$$X = \frac{V_\beta c_{\beta\alpha}}{V_0(\bar{c} - c_{\alpha\beta})} \tag{6.18}$$

where V_0 is the total volume of the system, V_β is the volume of β particles at time t, and the concentration terms have their usual meaning. Now the extended volume fraction is given by

$$X_{\text{ext}} = \frac{4\pi r^3 N_v c_{\beta\alpha}}{3(\bar{c} - c_{\alpha\beta})} \tag{6.19}$$

where r is the radius of the spherical precipitate particles and N_v is the number of such particles per unit volume.

Rewriting Eq. 6.19,

$$r = \left[\frac{3X_{\text{ext}}(\bar{c} - c_{\alpha\beta}^{\infty})}{4\pi N_v c_{\beta\alpha}}\right]^{1/3} \tag{6.20}$$

Also,

$$\frac{dX_{\text{ext}}}{dt} = 4\pi r^2 \frac{dr}{dt} \frac{N_v c_{\beta\alpha}}{(\bar{c} - c_{\alpha\beta}^{\infty})} \tag{6.21}$$

where

$$\frac{dr}{dt} = U = \frac{D(\bar{c} - c_{\alpha\beta}^{\infty})}{r(c_{\beta\alpha} - c_{\alpha\beta}^{\infty})} \qquad \text{for } r \gg r^* \tag{6.22}$$

Substituting for r and dr/dt in Eq. 6.21 from 6.20 and 6.22, we have

$$\frac{dX_{\text{ext}}}{dt} = \frac{4\pi D N_v c_{\beta\alpha}}{(c_{\beta\alpha} - c_{\alpha\beta}^{\infty})}\left[\frac{3X_{\text{ext}}(\bar{c} - c_{\alpha\beta}^{\infty})}{4\pi N_v c_{\beta\alpha}}\right]^{1/3}$$

$$= D N_v^{2/3} X_{\text{ext}}^{1/3} \kappa \tag{6.23}$$

where

$$\kappa = \left[\frac{48\pi^2 c_{\beta\alpha}^2(\bar{c} - c_{\alpha\beta}^{\infty})}{(c_{\beta\alpha} - c_{\alpha\beta}^{\infty})^3}\right]^{1/3} \tag{6.24}$$

Integrating Eq. 6.23, we obtain

$$\int_0^X \frac{dX_{\text{ext}}}{X_{\text{ext}}^{1/3}} = D N_v^{2/3} \kappa \int_0^t dt$$

or

$$X = \left(\frac{2}{3} D_\alpha N_v^{2/3} \kappa\right)^{3/2} t^{3/2} \tag{6.25}$$

Equation 6.25 is not corrected for impingement. At large volume fractions of the precipitate, a correction for impingement is necessary. Here, no physical impingement of the β particles occurs, as the depleted matrix always separates the β particles. However, the *diffusion fields of neighbouring β particles may overlap*, as illustrated for time t_2 in Fig. 6.6. The concentration in the matrix at large distances from the growing particles is no longer $\bar{c}$ but less than that. This overlap of diffusion fields is known as "soft impingement", whereas "hard impingement" refers to the physical impingement of the transformed particles, which occupy the entire volume on completion of the transformation, as discussed in Section 6.2.

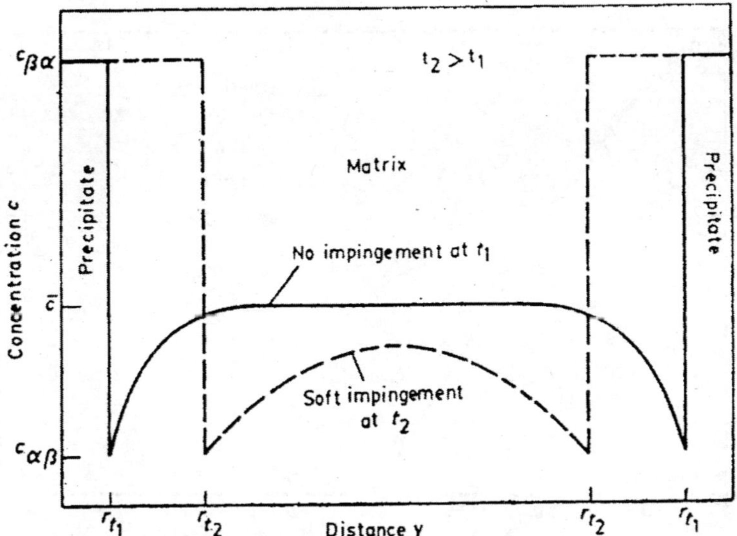

Fig. 6.6 During soft impingement, the diffusion fields of neighbouring precipitate particles overlap.

If we correct for impingement using the factor $(1 - X)$, we get

$$\frac{dX}{dt} = \frac{(1 - X)\, dX_{ext}}{dt} \tag{6.26}$$

This leads to the exponential form of the transformation kinetics (see Section 6.1):

$$X \approx 1 - \exp\left[-\left(\frac{2}{3}D_\alpha N_v^{2/3}\kappa\right)^{3/2} t^{3/2}\right] \tag{6.27}$$

the time exponent in both Eqs. 6.25 and 6.27 is 3/2 and the reaction rate constant $K = \frac{2}{3}D_\alpha N_v^{2/3}\kappa$.

Wert and Zener have followed the kinetics of continuous precipitation of spherical particles of cementite from supersaturated ferrite measuring the corresponding changes in the internal friction peak. Their data shown in Fig. 6.7 obey the transformation kinetics corrected for impingement (Eqs. 6.26 and 6.27). In Fig. 6.7, $1 - X$ is plotted against the product of the reaction rate constant and time, Kt, with K being evaluated experimentally at the four reaction temperatures between 86° and 211°C.

Ham gave a more general treatment of the transformation kinetics applicable to product particles of all shapes. He assumed (as in the above treatment) that a constant concentration of B atoms, $c_{\alpha\beta}$, exists in the α matrix that is in contact with the β particles. The β particles are taken to be distributed periodically in the matrix. Ham's results show that, irrespective of the shape, each dimension of the particle grows as the square root of time and the volume varies as 3/2, just as shown above for a sphere.

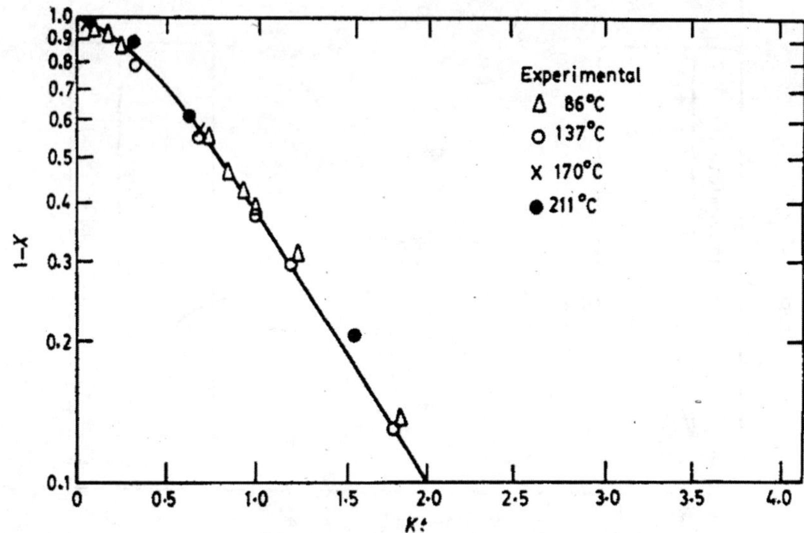

Fig. 6.7 The cementite precipitation data of Wert and Zener compared with the X versus t relationship given in Eq. 6.27.

Time Exponents in Diffusion-controlled Growth

For diffusion-controlled growth, a time exponent of $\frac{1}{2}$ arises for each *growing dimension* of the particle. For a constant rate of nucleation, an additional exponent of 1 arises. Table 6.2 summarises the time exponents for different situations.

TABLE 6.2

Time Exponents for Diffusion-controlled Growth

Description of the transformation	*Time exponent*
Particles of all shapes growing in all three dimensions	
1 increasing I	$> 2\frac{1}{2}$
2 constant I	$2\frac{1}{2}$
3 decreasing I	$1\frac{1}{2}$–$2\frac{1}{2}$
4 zero I	$1\frac{1}{2}$
Rod-like particles growing radially, zero I	1
Disc-like particle thickening, zero I	$\frac{1}{2}$

FURTHER READING

J.W. Christian, *The Theory of Transformations in Metals and Alloys*, Chapter 12, Formal Theory of Transformation Kinetics, Pergamon Press, Oxford, p. 525 (1975).

EXERCISES

6.1 In a diffusion-controlled transformation, the time t at different temperatures to yield the same fraction transformed X is as follows:

T°C	t, hr
25	24
100	1
150	0.25

Calculate the activation energy for the diffusion process by suitably modifying the X-t equation applicable here and plotting the data.

6.2 Using the exponential form of the empirical kinetic equation, determine the reaction-rate constant K and the time exponent n for the following transformation data:

Time, $t(s)$	X
3.5	0.046
4.8	0.19
6.3	0.31
6.9	0.49
7.6	0.60
9.5	0.80
10.0	0.93

Compute the rate of transformation at $X = 0.5$. Is this the maximum rate of transformation? Assuming that the Johnson-Mehl equation is valid for the above transformation and $U = 10^{-3}$ mm s^{-1}, calculate the nucleation rate.

6.3 With the data of Problem 6.2, calculate and plot the interfacial area per unit volume between parent and product phases (a) as a function of time, and (b) as a function of fraction transformed. What is the physical reason for the shape of the above plots?

6.4 The time exponent in the exponential form of the empirical equation for transformation kinetics is found to vary from 0.5 to 6 for different transformations. Discuss the factors that determine n and the experiments you would design to evaluate these factors.

7

Particle Coarsening

Coarsening is a structural change in which the average size of the precipitate particles in a system increases with time, usually after the precipitation process is complete. Ostwald ripening, coalescence and overageing are other terms which are commonly used to refer to this phenomenon. Coarsening is a process of practical importance, as the optimum mechanical properties of an alloy often depend on a critical size distribution of the precipitate particles which are in the submicroscopic range. Coarsening increases both the average size of the particles and the distance of separation between neighbouring particles, but it decreases the number of particles in the system. The stress required to move a dislocation through the matrix in which the precipitate particles are embedded is decreased as a result of coarsening thereby softening the alloy. The coarsening rate at the service temperature in such an alloy has to be sufficiently slow, so that no significant softening will occur due to coarsening within the life time of the structural component made from that alloy.

In this chapter, we first describe how the driving force for coarsening arises, using the Thomson-Freundlich equation. The simple theory of coarsening due to Greenwood is then described, with the derivation of the well-known cube-root relationship between the coarsening time and the average particle size.

7.1 DRIVING FORCE FOR COARSENING

The reduction in the total interfacial energy of the system provides the driving force for coarsening. The surface area-to-volume ratio of spherical particles varies inversely as their radius. So, for a given volume fraction of such particles, the total surface area decreases with an increase in the average particle size. Hence, coarsening decreases the total interfacial energy.

The Thomson-Freundlich Equation

The equation basic to the kinetic analysis of coarsening is the Thomson-Freundlich solubility relationship (also known as the Gibbs-Thomson equation):

$$c^{r}_{\alpha\beta} = c^{\infty}_{\alpha\beta} \exp\left(\frac{2V\sigma}{RTr}\right) \tag{7.1}$$

where

$c^{r}_{\alpha\beta}$ is the composition of the α matrix in equilibrium with a spherical particle β of radius r (or a particle with the local radius of curvature r),

$c^{\infty}_{\alpha\beta}$ is the matrix composition in equilibrium with a particle of very large radius,

V is the molar volume of the precipitate, and

σ is the specific interfacial energy.

This equation is applicable to an *ideal solution.*

Typical values of the parameters within the exponential in Eq. 7.1 are: $V = 7 \times 10^{-6}\ \text{m}^3$, $\sigma = 0.3\ \text{J m}^{-2}$, $r = 10^{-8}$ m and $T = 1000$ K. With these values, $2V\sigma/RTr \approx 0.05$. This can be taken to be small as compared with unity, so that Eq. 7.1 can be simplified to

$$c^{r}_{\alpha\beta} = c^{\infty}_{\alpha\beta}\left(1 + \frac{2V\sigma}{RTr}\right) \tag{7.2}$$

Equations 7.1 and 7.2 indicate that the solubility in the matrix in contact with a β particle increases, as the radius of the β particle decreases. Figure 7.1(a) shows schematically the free-energy/composition relationships. Here, the free energy of the β particle includes its surface energy. It increases with decreasing particle size. The common tangents drawn at the three particle sizes $r^* < r < \infty$ in Fig. 7.1(a) yield the matrix compositions as $c^{r^*}_{\alpha\beta} > c^{r}_{\alpha\beta} > c^{\infty}_{\alpha\beta}$. Figure 7.1(b) is the corresponding phase diagram. Note that the solubility in the α phase increases with decreasing precipitate size, compare with Fig. 1.3.

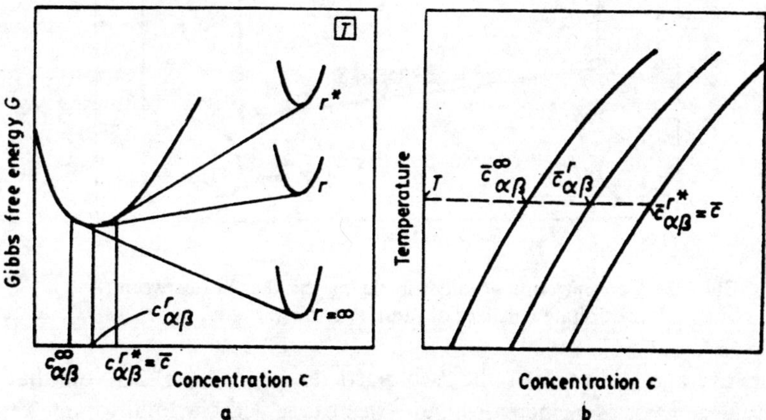

Fig. 7.1 (a) Schematic illustration of free energy-composition relationships for particles of varying sizes in equilibrium with the matrix. (b) The corresponding phase diagram.

Interaction between Two Particles

Consider the interaction between two spherical neighbouring precipitate particles embedded in the α matrix. Let their radii be r_1 and r_2. We can then write from Eq. 7.2:

$$c_{\alpha\beta}^{r_1} - c_{\alpha\beta}^{r_2} = c_{\alpha\beta}^{\infty} \frac{2V\sigma}{RT}\left(\frac{1}{r_1} - \frac{1}{r_2}\right) \tag{7.3}$$

For $r_1 < r_2$, we have $1/r_1 > 1/r_2$ and $c_{\alpha\beta}^{r_1} > c_{\alpha\beta}^{r_2}$. Thus a concentration gradient exists in the matrix between two particles of different sizes, if local equilibrium prevails near each particle. So, diffusional flow will occur *down* this concentration gradient, i.e., from the smaller particle to the larger particle, as schematically illustrated in Fig. 7.2. As the flow occurs, some part of the smaller particle dissolves to compensate for the decrease in concentration at the interface due to the flow. Likewise, some amount of the solute will precipitate at the bigger particle to remove the excess solute arriving as the diffusional flux. As this occurs, the difference in sizes of the two particles increases. Correspondingly, the concentration gradient becomes steeper. The process continues till the smaller particle is completely dissolved.

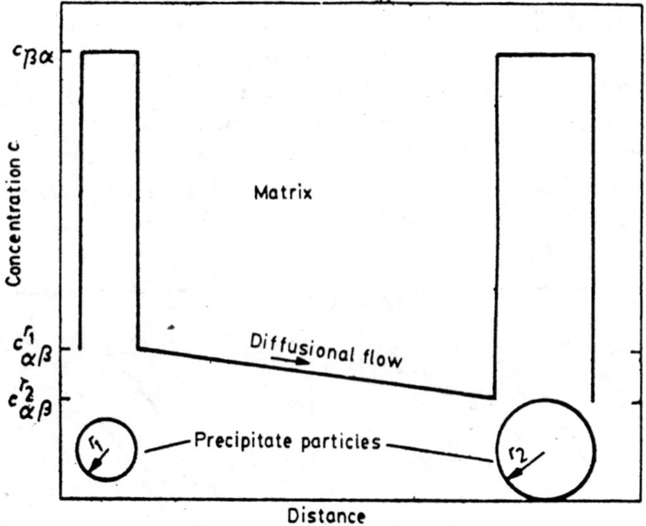

Fig. 7.2 Concentration gradient in the matrix that lies between two precipitate particles of radii, r_1 and $r_2 (r_1 < r_2)$.

The average radius $\bar{r}$ of the two particles is $(r_1 + r_2)/2$. Note that $\bar{r}$ actually decreases, as r_2 increases and r_1 decreases, with the total volume kept constant. But when r_1 becomes zero and hence is no longer counted in the average, $\bar{r}$ jumps from $(r_1 \to 0 + r_2)/2$ to r_2. Therefore, the coarsening is mainly due to the decrease in the number of particles.

7.2 THE KINETICS OF COARSENING (GREENWOOD'S MODEL)

General Features

A general kinetic treatment of particle coarsening must consider the following factors:

1 particles of arbitrary shapes
2 variation in the volume fraction of the precipitate during coarsening
3 anisotropy of surface energy
4 deviations from ideality of the matrix solution.

Further, the rate controlling step during coarsening may be either the long range diffusion of the solute through the matrix or its transfer across the interface. Note that Eq. 7.2 can be applied only when the process is not interface-controlled, as the equation assumes equilibrium between matrix and the particle at the interface. If diffusion through the matrix is rate controlling, the variation in the diffusion coefficient with concentration should also be considered. A rigorous treatment taking into account all these effects is extremely difficult. Here, a simple model due to Greenwood will be discussed.

Greenwood developed a model for the particle-coarsening kinetics based on the following assumptions:

1 the particles are spherical in shape

2 the distribution of the particles is random

3 the matrix is an ideal solid solution

4 the process is controlled by the diffusion of the solute atoms through the matrix

5 the diffusion coefficient of the solute is independent of the matrix composition

6 no fresh precipitation occurs during coarsening, i.e., the volume fraction of the precipitates remains constant.

Growth Rate of the ith Particle of Radius r_i

Equation 7.2 can be rewritten as:

$$c_{\alpha\beta}^{r_i} = c_{\alpha\beta}^{\infty}\left(1 + \frac{2V\sigma}{RTr_i}\right) \tag{7.4}$$

where r_i is the radius of the ith particle. From the general growth equation (Eq. 5.11) for diffusion-controlled processes, we can write:

$$U_i = \frac{dr_i}{dt} = \frac{D_\alpha\left(\frac{\partial c}{\partial y}\right)_{y=r_i}}{c_{\beta\alpha} - c_{\alpha\beta}^{r_i}} \tag{7.5}$$

It is now assumed that each particle sees a composition c' at large distances, where $c' > c_{\alpha\beta}^{\infty}$. This increase over the equilibrium solubility corresponding to an infinite radius of curvature arises, as all the other particles have finite radii of curvature. Following Eqs. 5.12 and 5.13, we can write

$$\left(\frac{\partial c}{\partial y}\right)_{y=r_i} = \frac{c' - c_{\alpha\beta}^{r_i}}{r_i} \tag{7.6}$$

Substituting Eq. 7.6 in 7.5, we obtain

$$U_i = \frac{D_\alpha}{r_i} \frac{c' - c_{\alpha\beta}^{r_i}}{c_{\beta\alpha} - c_{\alpha\beta}^{\infty}} \tag{7.7}$$

where $c_{\alpha\beta}^{r_i}$ in the denominator has been replaced by $c_{\alpha\beta}^{\infty}$, in comparison with $c_{\beta\alpha}$.

Let there be N_p particles at time t. Since

$$\sum_i^{N_p} V_i = \text{constant}, \tag{7.8}$$

we have

$$\sum_i^{N_p} \frac{dV_i}{dt} = 0 \tag{7.9}$$

As $V_i = \frac{4}{3}\pi r_i^3$ and $dV_i = 4\pi r_i^2\, dr_i$, we get

$$\sum_i^{N_p} 4\pi r_i^2 \frac{dr_i}{dt} = 0 \tag{7.10}$$

Substituting for $dr_i/dt = U_i$ from Eq. 7.7 in 7.10, we obtain

$$\sum_i^{N_p} 4\pi r_i^2 \frac{D_\alpha}{r_i} \frac{c' - c_{\alpha\beta}^{r_i}}{c_{\beta\alpha} - c_{\alpha\beta}^{\infty}} = 0 \tag{7.11}$$

The terms $4\pi D_\alpha/(c_{\beta\alpha} - c_{\alpha\beta}^{\infty})$ can be taken out of the summation. So,

$$\sum_i^{N_p} r_i(c' - c_{\alpha\beta}^{r_i}) = 0 \tag{7.12}$$

Using $c_{\alpha\beta}^{r_i} = c_{\alpha\beta}^{\infty}(1 + 2V\sigma/RTr_i)$ in Eq. 7.12,

$$\sum_i^{N_p} r_i\left(c' - c_{\alpha\beta}^{\infty} - c_{\alpha\beta}^{\infty}\frac{2V\sigma}{RTr_i}\right) = 0$$

or

$$\sum_i^{N_p} r_i(c' - c_{\alpha\beta}^{\infty}) = N_p c_{\alpha\beta}^{\infty}\frac{2V\sigma}{RT}$$

or

$$\frac{1}{N_p}\sum_i^{N_p} r_i = \bar{r} = \frac{c_{\alpha\beta}^{\infty} 2V\sigma}{(c' - c_{\alpha\beta}^{\infty})RT} \tag{7.13}$$

where $\bar{r}$ is the mean radius.

Rewriting Eq. 7.13, we have

$$c' = c_{\alpha\beta}^{\infty}\left(1 + \frac{2V\sigma}{RT\bar{r}}\right) \tag{7.14}$$

As this is identical in form to the Thomson-Freundlich equation, we conclude the $c' = c_{\alpha\beta}^{\bar{r}}$, i.e., the concentration that a particle sees at large distances is the same as the matrix concentration that is in equilibrium with a particle of mean radius $\bar{r}$.

The concentration profiles near particles of different sizes are shown schematically in Fig. 7.3. A particle of mean radius $\bar{r}$ sees no concentration gradient ahead of it. A particle of size $r_1 < \bar{r}$ sees a downward concentration gradient ahead of it, which favours its dissolution. A particle of size $r_2 > \bar{r}$ sees an upward concentration gradient ahead of it, which favours the arrivial of fresh solute atoms at its interface and growth of that particle. The diffusion directions are indicated by arrows in Fig. 7.3.

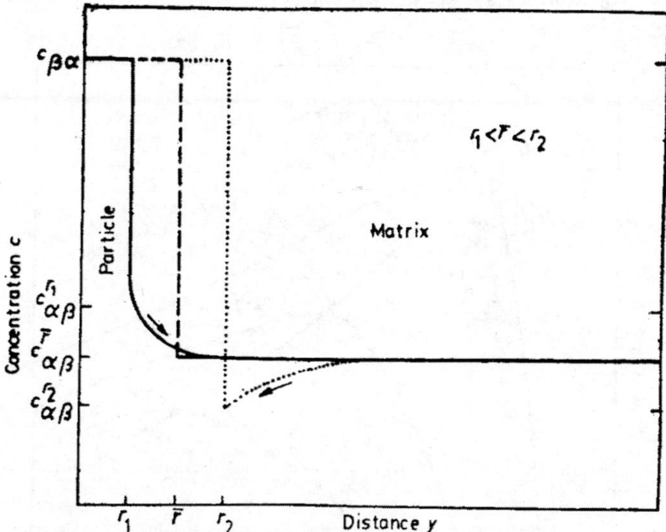

Fig. 7.3 Concentration gradient ahead of particles of size, $r_1 < \bar{r} < r_2$.

When the expression for c' (Eq. 7.14) is inserted in Eq. 7.7, we obtain

$$U_i = \frac{dr_i}{dt} = \frac{D_\alpha}{r_i}\,\frac{c_{\alpha\beta}^{\infty}}{c_{\beta\alpha} - c_{\alpha\beta}^{\infty}}\,\frac{2V\sigma}{RT}\left(\frac{1}{\bar{r}} - \frac{1}{r_i}\right) \tag{7.15}$$

This yields

1 $U_i < 0$ for $r_i < \bar{r}$
2 $U_i = 0$ for $r_i = \bar{r}$
3 $U_i > 0$ for $r_i > \bar{r}$

A negative growth rate indicates that particles with $r_i < \bar{r}$ will shrink. A positive U_i is indicative of the growth of the particle to larger sizes.

Fastest Growing Particles

The maximum growth rate U_{max} can be obtained from

$$\frac{dU_i}{dr_i} = \frac{1}{dr_i}\left[\frac{1}{r_i}\left(\frac{1}{\bar{r}} - \frac{1}{r_i}\right)\right] = 0 \tag{7.16}$$

Solving, we find that the particle that grows fastest in a given array of particles is that which has a radius $r_m = 2\bar{r}$. Substituting this in Eq. 7.15, we obtain

$$U_{max} = \frac{dr_m}{dt} = D_\alpha \frac{c^{\infty}_{\alpha\beta}}{c_{\beta\alpha} - c^{\infty}_{\alpha\beta}} \frac{2V\sigma}{RT} \frac{1}{r_m^2} \tag{7.17}$$

Thus the maximum growth rate is inversely proportional to r_m^2 or $\bar{r}^2$. The variation in growth rate as a function of the mean particle size at different times of the coarsening anneal are shown in Fig. 7.4. The locus of the radius of the fastest growing particle is also shown.

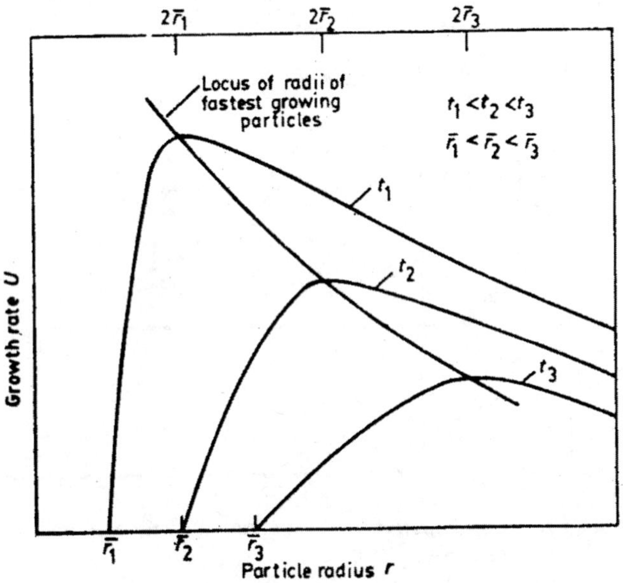

Fig. 7.4 Growth rate of particles as a function of their size, with increasing coarsening times, $t_3 > t_2 > t_1$.

The Cube-root Relationship

The rate of coarsening can be expressed as a functional relationship between $\bar{r}$ and t. Consider particles of radius $r_m = 2\bar{r}$. Integrating Eq. 7.17, we obtain

$$\int_{r_m^0}^{r_m} r_m^2 dr_m = D_\alpha \frac{c^{\infty}_{\alpha\beta}}{c_{\beta\alpha} - c^{\infty}_{\alpha\beta}} \frac{2V\sigma}{RT} \int_0^t dt \tag{7.18}$$

where $\overset{0}{r_m} = 2\bar{r}^0$ at the beginning of the coarsening process. Solving for $\bar{r}$, we get

$$\bar{r} = \left[(\bar{r}^0)^3 + \frac{3}{8} D_\alpha \frac{c_{\alpha\beta}^{\infty}}{c_{\beta\alpha} - c_{\alpha\beta}^{\infty}} \frac{2V\sigma}{RT} t\right]^{1/3} \tag{7.19}$$

The mean radius $\bar{r}$ can be determined by quantitative metallography from the Fullman's relationship:

$$\bar{r} = \frac{\pi}{4\bar{M}} \tag{7.20}$$

where $\bar{M}$ is the mean of the reciprocals of the diameters of the circular intersections of the particles with a random test plane.

Experimental data can be checked for the above cube-root relationship as follows. If $\bar{r}^0$ is very small as compared to $\bar{r}$, a plot of $\bar{r}$ against $t^{1/3}$ should yield a straight line. Such a plot for the coarsening of copper precipitates in an iron matrix is shown in Fig. 7.5. If $\bar{r}^0$ is not small enough to be neglected,

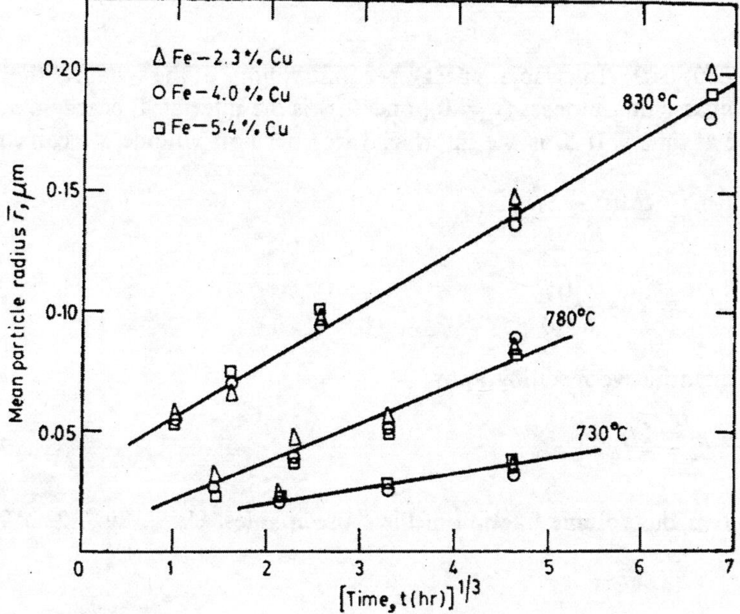

Fig. 7.5 $\bar{r}$ versus $t^{1/3}$ plot for the coarsening of copper precipitates in an iron matrix.

then a plot of $\bar{r}^3$ against t should yield a straight line. Figure 7.6 shows such a plot for the coarsening of $Mg_2Al_3(\beta)$ precipitates in an aluminium matrix.

An alternative way of expressing the coarsening kinetics is to plot X versus t, where X is the fractional decrease in the interfacial energy:

$$X = \frac{U(0) - U(t)}{U(0)} \tag{7.21}$$

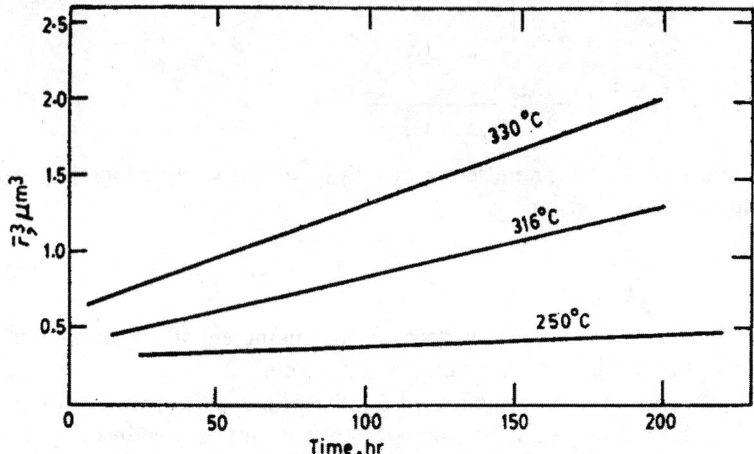

Fig. 7.6 $\bar{r}^3$ versus t plot for the coarsening of Mg_2Al_3 (β) precipitates in an aluminium matrix.

where $U(0)$ is the interfacial energy per unit volume of the system at the start of the coarsening process ($t = 0$), and $U(t)$ is the interfacial energy per unit volume at time t. If S_v is the interfacial area per unit volume, we can write

$$X = \frac{S_v(0) - S_v(t)}{S_v(0)}$$

$$= 1 - \frac{S_v(t)}{S_v(0)} \tag{7.22}$$

From quantitative metallography,

$$S_v = \frac{2f_\beta}{\bar{r}} \tag{7.23}$$

where f_β is the volume fraction of the β precipitates. Using Eq. 7.23 in 7.22, we get

$$X = 1 - \frac{\bar{r}^0}{\bar{r}(t)} \tag{7.24}$$

Substituting for $\bar{r}(t)$ from Eq. 7.19 in 7.24, we have

$$X = 1 - (1 + Kt)^{-1/3} \tag{7.25}$$

where

$$K = \frac{3}{8} \frac{D_\alpha}{\bar{r}^{03}} \frac{c^\infty_{\alpha\beta}}{c_{\beta\alpha} - c^\infty_{\alpha\beta}} \frac{2V\sigma}{RT}$$

FURTHER READING

G.W. Greenwood, *Acta Metall.*, **4**, 243 (1956).

EXERCISE

7.1 Calculate and plot the composition of ferrite coexisting with cementite spheres in the temperature range of 550 to 725°C, when the cementite particles have a radius of (i) 10^{-5} cm, (ii) 10^{-6} cm, and (iii) 10^{-7} cm. Given: $\ln c_{\alpha-Fe_3C} = 0.936 - 4350/T$, where c is wt. % carbon in ferrite, when the cementite particles are very large. $\sigma = 0.7$ J m^{-2}; and density of cementite = 7700 kg m^{-3}.

8

Pearlitic Transformations

The eutectoidal transformation in iron-carbon alloys and other alloy steels has been studied extensively, as it is a transformation of great importance in steel technology. Though the discussion in this chapter is more specifically on the austenite-pearlite eutectoidal transformation in steels, the underlying principles have wider applicability in discontinuous precipitation reactions occurring in other systems. For example, the cellular precipitation of tin from a lead-tin solid solution has characteristics quite similar to the pearlitic transformation in steels.

In this chapter, the distinction between continuous and discontinuous precipitation is first brought out. After describing the experimental characteristics of pearlitic transformations, the mechanism and the kinetics of nucleation and growth of pearlite are discussed in some detail. Finally, aspects about the interlamellar spacing and the effect of alloying elements on the transformation are described.

8.1 CONTINUOUS VERSUS DISCONTINUOUS PRECIPITATION

In continuous precipitation, particles of the product β phase rich in solute nucleate and grow with a *continuous* decrease in solute concentration of the supersaturated matrix α, by solute diffusion across a concentration gradient ahead of the growing particle:

$$\alpha_{\text{supersat}} \rightarrow \alpha_{\text{sat}} + \beta \tag{8.1}$$

The parent α phase, except for the decrease in the solute concentration, retains its crystal structure and *matrix orientation*. The kinetic equations developed in Section 6.3 are applicable to this type of precipitation.

In *discontinuous precipitation*, on the other hand, cells (or colonies) of lamellae or rods lying generally parallel to one another within a given cell (or colony), precipitate from a supersaturated solution:

$$\alpha_{\text{supersat}} \rightarrow \beta + \gamma \tag{8.2}$$

The average composition of the two (or sometimes more) product phases ($\beta + \gamma$) *within each cell* is equal to the overall composition of the parent phase α. As the transformation proceeds, product phases advance sideways and lengthwise. The cellular configuration grows at the expense of the parent phase, with local changes in the matrix concentration near the interface. The sidewise and the lengthwise growth of a colony of parallel plates is schematically illustrated in Fig. 8.1.

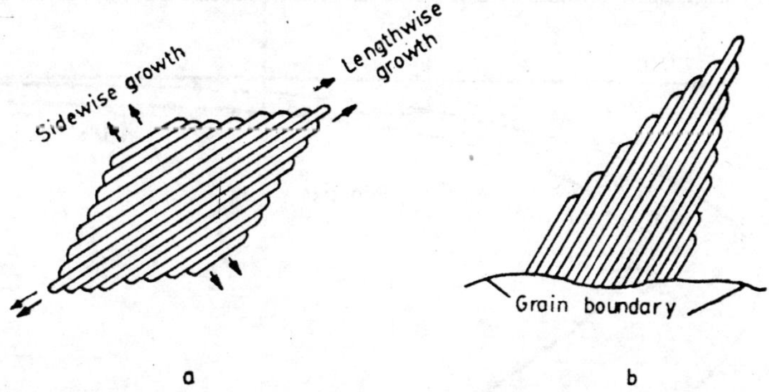

Fig. 8.1 Sidewise and lengthwise growth of a pearlite colony.

A typical example of the discontinuous precipitation process is the eutectoidal decomposition of austenite (γ) in steels to pearlite, which is a mixture of lamellar ferrite (α) and cementite (Fe_3C). Another example is the tin-rich precipitate from a lead-tin solid solution, with a simultaneous *reorientation* of the solid solution matrix. In this case, the reaction is similar to that given in Eq. 8.1, but with the important difference that the transformation is accompanied by a reorientation of the matrix and occurs in cellular morphology.

8.2 EXPERIMENTAL CHARACTERISTICS

The transformation of austenite to pearlite is usually depicted in the form of an isothermal transformation (I-T) diagram, known also as a time-temperature-transformation (T-T-T) diagram. Figure 8.2 is such a diagram for a 0.8% carbon steel. Two C-curves characterize the start and the finish of the transformation of austenite to pearlite or bainite. The upper part of the C-curve that lies below the eutectoid temperature of 727°C is the pearlitic region. The nose of the C-curve coincides with the temperature of the maximum rate of transformation. The time interval for the transformation to start at the nose temperature determines the ease with which the pearlitic transformation can be suppressed during hardening of the steel. This time increases with increasing austenitic grain size and also in the presence of common alloying elements such as Cr, Mn, Mo, and Ni.

The nucleation of pearlite occurs predominantly at the austenite grain boundaries. The finer is the grain size of austenite, the more is the boundary area per unit volume and also the number of potential nucleation sites. This explains the higher transformation rate observed in finer-grained austenite. Figure 8.3 depicts the sigmoidal transformation curves for three different

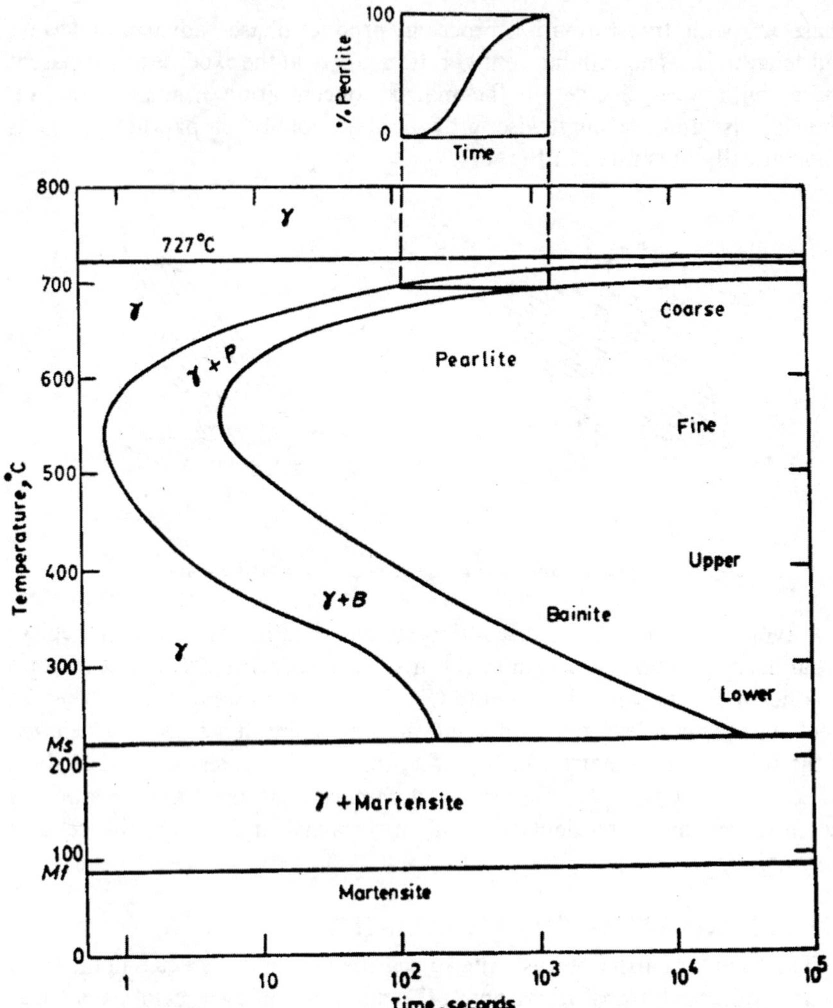

Fig. 8.2 Isothermal transformation (I–T) diagram for a 0.8% carbon steel.

grain sizes of austenite. Also, the predominant boundary nucleation results in site saturation in the early stages of the transformation and slab-like growth from the boundaries towards the centre of the grains, refer to Section 6.2. The time dependence of the nucleation rate in the early stages shows that the nucleation rate increases as the square of time. The rate of nucleation increases with decreasing temperature of the reaction down to 550°C, as illustrated in Fig. 8.4.

The colonies or nodules of pearlite often develop on only one side of a grain boundary, see Fig. 8.1(b). Hillert showed that the morphology of pearlite can be explained by a branching mechanism. He demonstrated by a successive etching technique that the carbide lamellae of a colony were completely interconnected. However, branching is not always observed in cellular reactions.

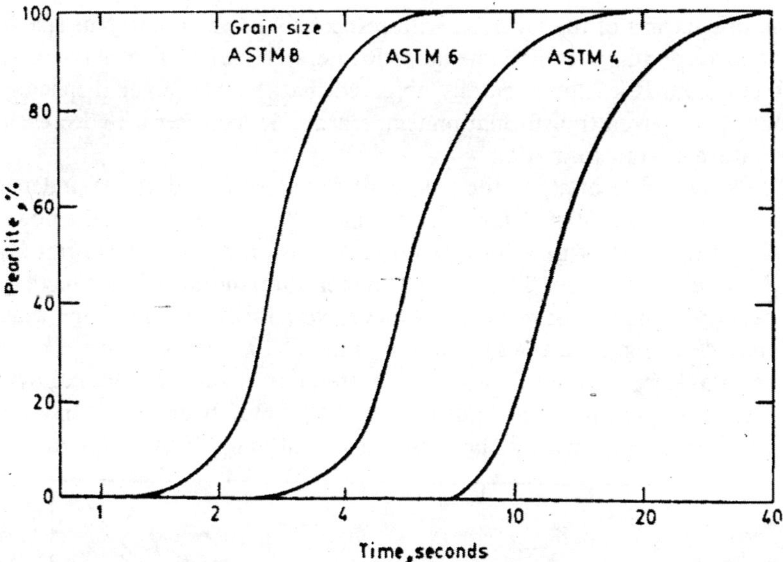

Fig. 8.3 The pearlite transformation kinetics for three austenite grain sizes.

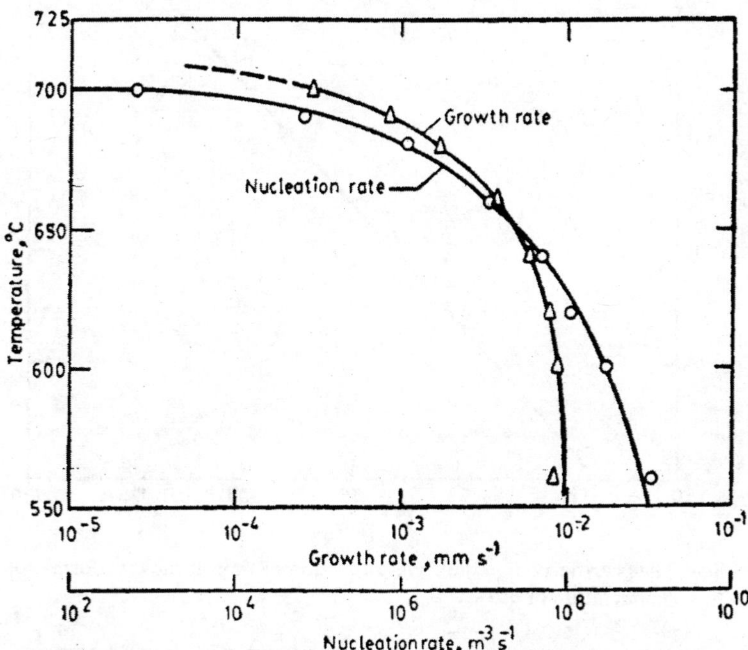

Fig. 8.4 The nucleation and the growth rates of pearlite as a function of temperature.

The interlamellar spacing in pearlite is defined as the distance from the centre of a cementite (or ferrite) plate to that of the next cementite (or ferrite) plate of a colony. On a random cross-sectional plane, this distance depends on the orientation of the lamellae with respect to the plane. The true spacing is that corresponding to the lamellae that intersect at right angles the plane of observation. It is experimentally observed that the interlamellar spacing is constant for a given transformation temperature. It decreases with decreasing temperature of transformation.

The growth rate of a pearlite colony or nodule is found to be independent of time at a constant temperature. Figure 8.5 shows the pearlite nodule radius as a function of time for a 0.8% C steel isothermally transformed at 680°C. On cooling below the eutectoid temperature, the growth rate increases first, reaches a maximum and then decreases with further fall in temperature, see Fig. 8.4. The growth rates are in the range of 10^{-4} – 10^{-2} mm s^{-1}. The growth rate decreases considerably in the presence of substitutional alloying elements. For example, the addition of 0.5% molybdenum to the steel decreases the growth rate by about two orders of magnitude.

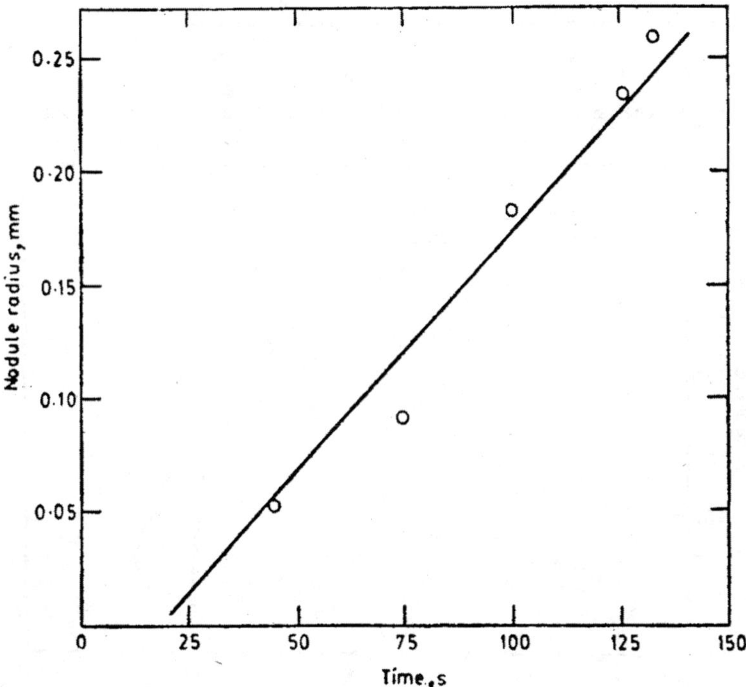

Fig. 8.5 The pearlite nodule radius as a function of time at the transformation temperature of 680°C.

8.3 THE MECHANISM AND THE KINETICS OF NUCLEATION

The Smith Hypothesis

Smith proposed that the lamellar reaction is started by the formation of a nucleus at the austenite grain boundary in an orientation which gives it a low-energy (coherent or semicoherent) interface with a grain on one side of the boundary. This results in an incoherent interface between the nucleus and the grain on the other side of the boundary, facilitating strain-free growth into the second grain. The incoherent boundary also provides a diffusion short-circuit for long range diffusion of the solute atoms that may be necessary to bring about the compositional changes of the reaction. The morphological observations such as the hemispherical growth of nodules of pearlite (Fig. 8.1(b)) and the crystallographic orientation relationships observed in some cellular reactions support this hypothesis.

The Time Dependence of the Nucleation Rate

Fisher has provided a simple picture for the development of a lamellar configuration. The nucleation and growth of either phase in the transformation product alter the local composition of the surrounding matrix in such a way as to favour the nucleation and growth of the other phase. For example, nucleation of a cementite particle depletes the surrounding matrix of carbon, which promotes the nucleation of ferrite. Thus, nucleation is a two-stage process.

Although this does not affect the subsequent argument, Fisher assumes that cementite nucleates first on an austenite grain boundary. The ferrite then nucleates at the cementite-austenite interface. The interfacial area between cementite and austenite increases, as the cementite particle grows. The probability of ferrite nucleation on this interface also increases. Once the first ferrite nucleus forms, the pearlite nucleation is taken to be complete.

In developing a model along these lines, Fisher makes the following assumptions:

1 The cementite particle is spherical in shape in the initial stages.

2 Its growth in the early stages is diffusion-controlled and is, therefore, parabolic.

3 The nucleation rate of cementite per unit area of the austenite grain boundary I_{cm} is constant at constant temperature.

4 The nucleation rate of ferrite per unit area of the austenite-cementite interface I_f is constant at constant temperature.

The surface area A of a cementite particle of radius r is

$$A = 4\pi r^2 \tag{8.3}$$

The initial parabolic growth of the particle yields

$$r = \alpha t^{1/2} \tag{8.4}$$

where α is a constant. Substituting Eq. 8.4 in 8.3, we get

$$A = 4\pi\alpha^2 t \tag{8.5}$$

If I'_f is the rate of ferrite nucleation per cementite particle,

$$\begin{aligned} I'_f &= I_f A \\ &= I_f 4\pi\alpha^2 t \end{aligned} \tag{8.6}$$

Let P be the probability that the first ferrite nucleus has formed on the cementite-austenite interface. This is the same as the probability that the pearlite has nucleated. $(1 - P)$ represents the probability that the first ferrite nucleus has *not* formed. We can write

$$dP = (1 - P)I_f' \, dt \tag{8.7}$$

Substituting for I_f' from Eq. 8.6 in 8.7, we obtain

$$dP = (1 - P)I_f 4\pi\alpha^2 t \, dt \tag{8.8}$$

Rearranging and integrating, we have

$$-\ln(1 - P) = I_f 2\pi\alpha^2 t^2$$

or

$$P = 1 - \exp(-mt^2) \tag{8.9}$$

where $m = I_f 2\pi\alpha^2$. From Eq. 8.9, we have

1 for $t = 0$, $P = 0$;
2 for $t = \infty$, $P = 1$.

If I_p' is the rate of pearlite nucleation per cementite particle, then

$$I_p' = \frac{dP}{dt}$$

$$= 2mt \exp(-mt^2) \tag{8.10}$$

The rate of pearlite nucleation per unit area of the austenite grain boundary I_p is obtained by integrating Eq. 8.10 to include all cementite particles that nucleate per unit area up to time t:

$$I_p = \int_{\tau=0}^{\tau=t} 2m(t - \tau) \exp[-m(t - \tau)^2] I_{cm} \, d\tau \tag{8.11}$$

This yields

$$I_p = I_{cm}[1 - \exp(-mt^2)] \tag{8.12}$$

For small values of mt^2,

$$I_p = I_{cm} mt^2 \tag{8.13}$$

Thus the pearlite nucleation rate increases as the square of time in the early stages of the reaction. Figure 8.6 compares experimental measurements of nucleation rates with those calculated from Eq. 8.13.

8.4 THE MECHANISM AND THE KINETICS OF GROWTH

Long-range Diffusion of Carbon

In plain carbon steels, the diffusion of carbon to effect the compositional changes necessary for the transformation is the rate controlling step. In an

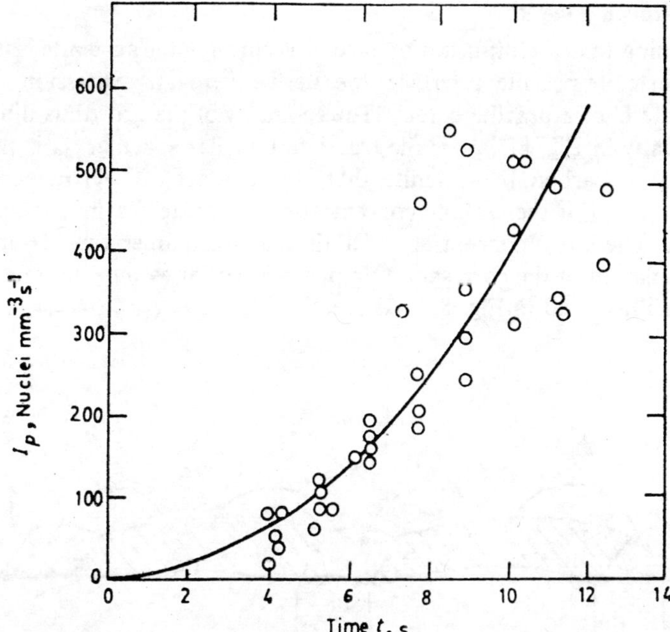

Fig. 8.6 The pearlite nucleation rate I_p as a function of time t in the early stages.

alloy steel, if the alloying element partitions itself in different amounts between ferrite and cementite, then the diffusion of the alloying element will be rate controlling, inasmuch as the substitutional alloying element diffuses much more slowly than interstitial carbon. If the alloying element does not redistribute during the transformation, then the diffusion of carbon will be rate controlling in the alloy steel as well. Here, we discuss the growth process controlled by the diffusion of carbon.

The diffusion of carbon can be through austenite, ferrite or the austenite-pearlite interface. By virtue of the greater solubility of carbon in austenite, larger concentration gradients are possible in austenite than in ferrite. On the other hand, at the temperatures of interest in pearlitic transformations, the diffusion coefficient for carbon in ferrite D_α is larger than the coefficient in austenite D_γ. The diffusion coefficient D_b for diffusion along the austenite-ferrite interface will be the largest. However, the cross-sectional area for diffusional flux is very much smaller here. A ratio of D_b/D_l (where l stands for the lattice) greater than 10^3 will be required to make the boundary diffusion dominant. Such a large ratio is not expected in the case of interstitial diffusion.

The carbon concentration gradients can exist parallel to the austenite-pearlite interface as well as normal to it. Even though solutions to the diffusion equation taking into account concentration gradients in more than one direction are available; we will consider only the concentration gradient parallel to the interface.

The Growth Kinetics

Assuming that the diffusion of carbon is through the austenite just ahead of the austenite-pearlite interface, we derive here an expression for the growth rate U of a pearlite colony. The geometry of the growth is illustrated schematically in Fig. 8.7. S_0 is the true interlamellar spacing. $c_{\gamma\alpha}$ is the concentration of carbon in austenite that is in contact with ferrite across the interface. $c_{\gamma-Fe_3C}$ is the carbon concentration in austenite in contact with cementite. These two concentrations at the reaction temperature are obtained by extrapolation of the corresponding phase boundaries into the metastable region, as illustrated in Fig. 8.8. As seen in the figure, $c_{\gamma\alpha} > c_{\gamma-Fe_3C}$, so that

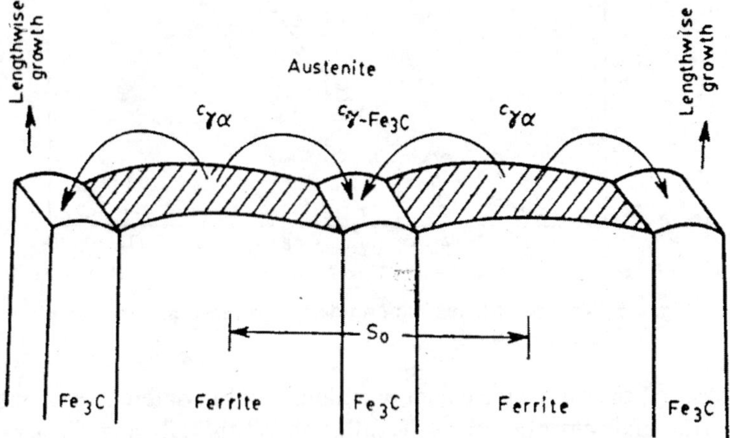

Fig. 8.7 The geometry of the lengthwise pearlitic growth under a concentration gradient in the austenite ahead of the growing colony.

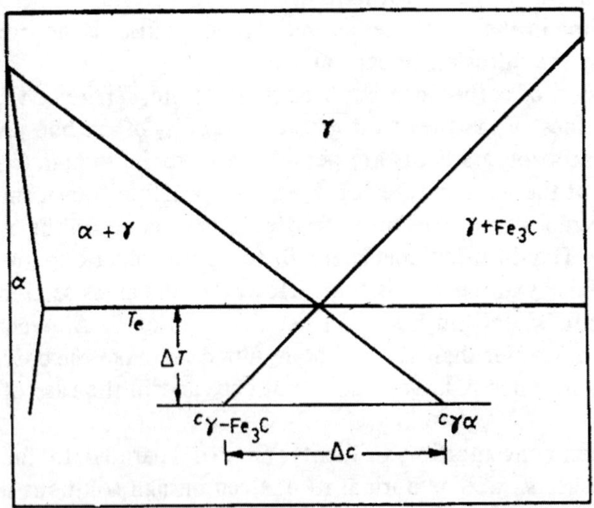

Fig. 8.8 Hultgren's extrapolation of the phase boundaries into the metastable region of an Fe–C phase diagram.

a concentration gradient exists in the austenite for carbon diffusion from regions which would transform to ferrite to regions that would form cementite. The directions of carbon diffusion are shown by arrows in Fig. 8.7.

The general growth equation for diffusion-controlled growth is given by Eq. 5.11:

$$U = \frac{D_\alpha}{c_{\beta\alpha} - c_{\alpha\beta}}\left(\frac{\partial c}{\partial y}\right)_{y=r}$$

For the problem at hand,

$$D_\alpha = D_c^\gamma$$

$$c_{\alpha\beta} = c_{\alpha-\mathrm{Fe_3C}}$$

and

$$c_{\beta\alpha} = c_{\mathrm{Fe_3C}} \tag{8.14}$$

$(\partial c/\partial y)_{y=r}$ is obtained from Hultgren's extrapolation referred to above, see Fig. 8.8. Lengthwise growth of the lamellae and a constant carbon concentration gradient in the austenite ahead of the interface are assumed. The latter assumption is valid, as the pearlitic transformation is a discontinuous process. From Figs. 8.7 and 8.8,

$$\Delta c = c_{\gamma\alpha} - c_{\gamma-\mathrm{Fe_3C}} \tag{8.15}$$

$$\Delta y = \tfrac{1}{2}S_0 \tag{8.16}$$

Then,

$$U = \frac{D_c^\gamma}{c_{\mathrm{Fe_3C}} - c_{\alpha-\mathrm{Fe_3C}}} \frac{2(c_{\gamma\alpha} - c_{\gamma-\mathrm{Fe_3C}})}{S_0} \tag{8.17}$$

If the extrapolated phase boundaries in Fig. 8.8 are taken to be straight lines, we can write

$$c_{\gamma\alpha} - c_{\alpha-\mathrm{Fe_3C}} \propto \Delta T \tag{8.18}$$

where ΔT is the degree of supercooling below the eutectoid temperature. Also, from Eq. 4.5, $|\Delta g| \propto \Delta T$. By substituting this and Eq. 8.18 in 8.17, we have

$$U = -\Gamma \frac{D_c^\gamma \Delta g}{S_0} \tag{8.19}$$

where Γ is a constant independent of time and temperature. The growth rate is, therefore, constant at constant temperature as found experimentally in pearlitic and other cellular transformations.

The interlamellar spacing S_0 is found to decrease linearly with temperature.

$$S_0 \propto \frac{1}{\Delta T} \tag{8.20}$$

So, we note from Eqs. 8.20, 8.19 and 8.18,

$$U \propto \exp\left(-\frac{Q}{RT}\right)(\Delta T)^2 \tag{8.21}$$

where Q is the activation energy for carbon diffusion in austenite. Accordingly, the growth rate is zero at $T = T_0$. It increases approximately as $(\Delta T)^2$, just below the eutectoid temperature. After reaching a maximum, the rate decreases again, as the exponential term in Eq. 8.21 dominates. Figure 8.4 compares the experimental measurements of growth rates as a function of temperature with those obtained from Eq. 8.21.

8.5 INTERLAMELLAR SPACING

In the previous section on growth kinetics, it was pointed out that the diffusion of carbon in austenite can be rate controlling. The diffusion distance is approximately half of the interlamellar spacing. As the spacing increases, the diffusion distance also increases and the concentration gradient decreases, resulting in a decrease in the growth rate. If the spacing is small, the growth rate is not necessarily large. With a decrease in spacing, the area of the α-Fe_3C interface per unit volume of pearlite increases. So the net driving force for the reaction decreases.

A net driving force Δg_{net} for the transformation is defined by taking into account the energy of the α-Fe_3C interfaces that lie within a pearlite colony:

$$\Delta g_{net} = \Delta g + A\sigma \tag{8.22}$$

where A is the area of the ferrite-cementite interfaces per unit volume of pearlite and σ is its specific surface energy. In a unit cube of pearlite, as we traverse in a direction perpendicular to the α-Fe_3C interfaces, for every length S_0, we cross two interfaces of unit area each. So, we have

$$A = \frac{2}{S_0} \tag{8.23}$$

Substituting for A from Eq. 8.23 in 8.22, we have

$$\Delta g_{net} = \Delta g + \frac{2\sigma}{S_0} \tag{8.24}$$

The minimum possible interlamellar spacing S_{min} corresponds to the case where the net driving force is zero.

$$\Delta g = -\frac{2\sigma}{S_{min}} \tag{8.25}$$

Substituting for Δg from Eq. 8.25 in 8.24, we have

$$\Delta g_{net} = -2\sigma\left(\frac{1}{S_{min}} - \frac{1}{S_0}\right) \tag{8.26}$$

The optimum value of S_0 corresponds to the maximum growth rate U_{max}. Substituting for Δg from Eq. 8.26 in 8.19, we have for the growth rate:

$$U = 2\sigma\Gamma\frac{D_c^{\gamma}}{S_0}\left(\frac{1}{S_{min}} - \frac{1}{S_0}\right) \tag{8.27}$$

Setting $dU/dS_0 = 0$, the condition for the maximum growth rate is obtained:

$$-\frac{1}{S_0^2 S_{min}} + \frac{2}{S_0^3} = 0$$

or

$$S_0 = 2S_{min} \tag{8.28}$$

The interlamellar spacing corresponding to the maximum growth rate is thus twice the minimum interlamellar spacing corresponding to the net driving force becoming zero. Since $|\Delta g| \propto \Delta T$ and $S_{min} = \frac{1}{2}S_0$, we see from Eq. 8.25 that S_0 should vary inversely as the degree of supercooling. Figure 8.9 is a log-log plot of the experimental values of interlamellar spacing S_0 as a function of ΔT. The slope is seen to be -1, indicating an inverse linear relationship between S_0 and ΔT. The ferrite-cementite interfacial energy σ can be evaluated from this plot and Eq. 8.27, and it turns to be about 3 J m^{-2}. This value is about an order of magnitude higher than that expected for a semicoherent interface.

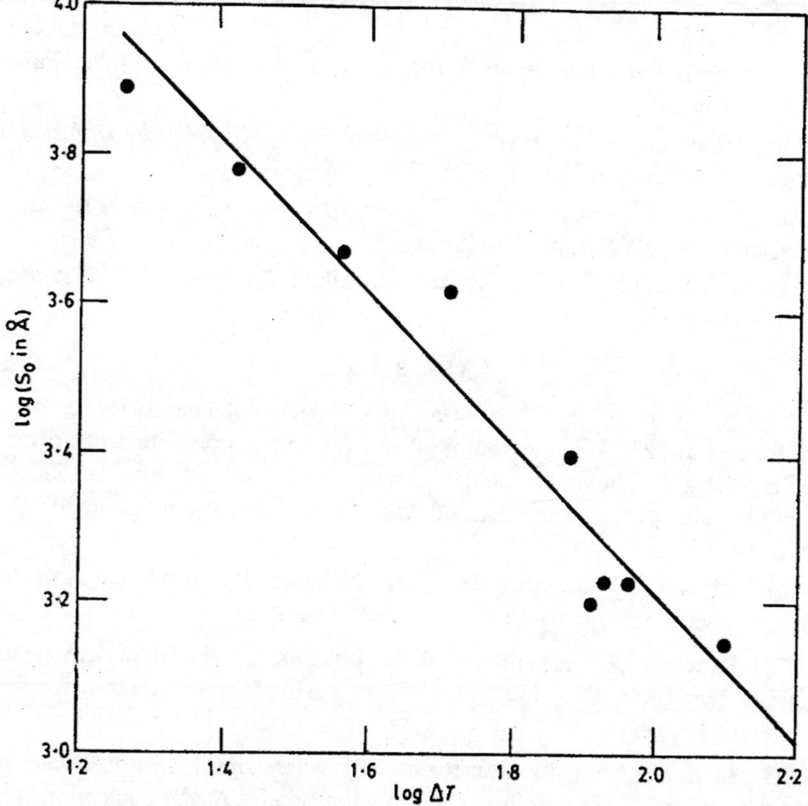

Fig. 8.9 The interlamellar spacing S_0 as a function of the degree of supercooling ΔT.

8.6 EFFECT OF ALLOYING ELEMENTS

The free energy change Δg for the austenite-pearlite reaction is altered by the presence of alloying elements in the steel. Most alloying elements decrease $|\Delta g|$, whereas cobalt increases $|\Delta g|$ at a given reaction temperature. The nucleation rate, the growth rate and the interlamellar spacing undergo corresponding changes.

The kinetic effects of the alloying elements are also quite marked. They influence the diffusion coefficient of carbon in austenite, although the effect is usually small. If the alloying element partitions, the slower diffusion rate of the substitutional atoms will determine the growth kinetics. The ratio of D_b/D_l, where b and l stand for boundary and lattice diffusion, is expected to be much larger than 10^3 for substitutional diffusion. Correspondingly, the diffusion short circuit provided by the austenite-pearlite interface will be operative in the presence of alloying elements. If the alloying element does not partition, the available driving force for the reaction is smaller than that for the case of equilibrium partitioning of the alloying element between ferrite and cementite.

FURTHER READING

J.C. Fisher, *Eutectoid Decompositions*, Amer. Soc. Metals, Metals Park, OH, p. 201 (1950).

J.W. Cahn and W.C. Hagel, *Decomposition of Austenite by Diffusional Processes*, Interscience, New York, p. 131 (1962).

M. Hillert, *Decomposition of Austenite by Diffusional Processes*, Interscience, New York, p. 197 (1962).

Symposium Papers on Cellular and Pearlite Reactions, *Metall. Trans.*, 3, 2717-2804 (1972).

EXERCISES

8.1 (a) Estimate the ferrite-cementite interfacial energy for a steel transformed at 623°C. The enthalpy of the austenite-to-pearlite transformation is 84 J g^{-1}. The interlamellar spacing at this temperature is 1500 Å. From the value you obtain, what do you deduce about the nature of the interface?

(b) Estimate the growth rate U for the same steel at 623°C. Take the diffusion coefficient of carbon $D_c^{\gamma} = 3 \times 10^{-11}$ m^2 s^{-1}.

8.2 How do you explain the wide variation in the time exponents obtained from plots of log log $(1/1 - X)$ versus log t for pearlite transformed at different temperatures?

8.3 In cases of pearlitic transformations, where site saturation occurs in the early stages of the reaction, the time dependence of the nucleation rate does not influence the time exponents obtained from the transformation kinetics. Explain why this is so.

9

Massive Transformations

In recent decades, it has been established that, in a number of alloy systems, the parent phase can transform to the product phase by short-range diffusion of atoms, that involves a relatively rapid uncoordinated transfer of atoms across the interface, without a change in composition. Such transformations are now recognized as a separate class of solid state reactions known as massive transformations.

In this chapter, we outline the experimental characteristics of massive transformations, comparing the massive growth rates with those of other transformations. The mechanism and the kinetics of the massive growth process are outlined in some detail.

9.1 EXPERIMENTAL CHARACTERISTICS

Massive transformations are characterized by no compositional change (like martensitic transformations) and by no shear displacements (unlike martensitic transformations). In 1930, Philips was the first to observe that, in Cu–Zn alloys, drastic quenching converted the high temperature β phase into "large units" of the α phase, without a change in composition. Ten years later, Greninger used the term "massive" structure to denote the blocky appearance of FCC α crystals obtained by quenching the high temperature β phase in the Cu–Al system. Knowledge concerning massive transformations has increased substantially in recent years. This type of transformation has now been identified in a number of alloys, such as Ag–Cd, Cu–Ga, Fe–C, Fe–Cr and Fe–Ni.

In order to induce a massive transformation in an alloy during continuous cooling, other competing reactions must be avoided. The rate of cooling should be in the right range. If the cooling rate is too slow, a competing reaction involving a compositional change can occur by means of long range diffusion. Too fast cooling rates, on the other hand, can induce a diffusionless martensitic transformation by shear displacement. A transformation temperature versus cooling rate plot in Fig. 9.1 for pure iron and Fe–Ni alloys exhibits two plateau regions. For any composition, the first plateau at the higher transformation temperature corresponds to the massive transformation. The second plateau at the lower temperature signifies the onset of the martensitic transformation. In Fe–10 at.% Ni alloy, there is only one plateau corresponding to the martensitic transformation. It is seen that there is a range of cooling rates over which the massive transformation can occur. The upper limit of this range decreases with increasing nickel content, from about 30,000°C s^{-1} for pure iron to 15,000°C s^{-1} for a Fe–7 at.% Ni alloy.

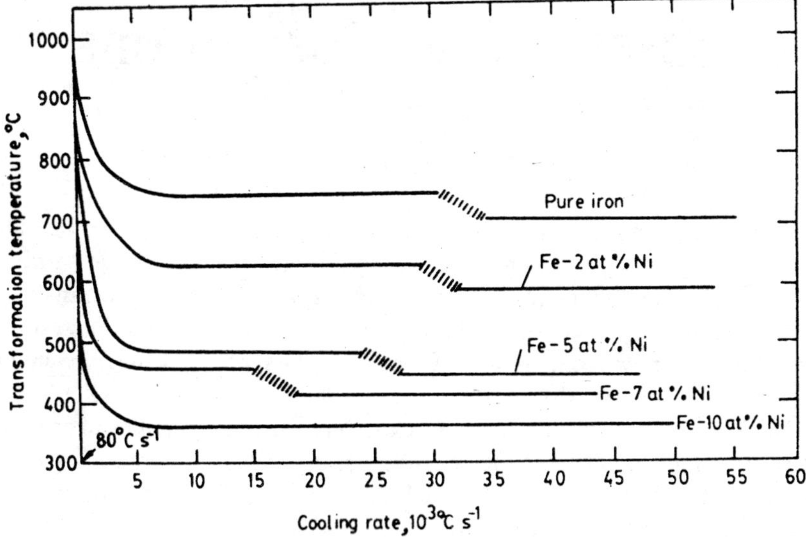

Fig. 9.1 Massive and martensitic transformation temperatures as a function of cooling rate in Fe-Ni alloys.

Nucleation of a massive reaction usually occurs at the grain boundaries of the parent phase, Fig. 9.2. Except in some special cases, there is usually

Fig. 9.2 Nucleation of the massive product at the parent grain boundaries. (Courtesy: The Institute of Metals, London.)

no particular orientation relationship between the product and the parent phases. The massive product grows roughly with a spherical front with irregular interfaces and may sweep across the grain boundaries of the parent phase.

The growth rate of the massive product is quite high, about 10 mm s^{-1}. This is compared with the growth rates observed in other transformations in Table 9.1. A growth rate of 10 mm s^{-1} means that the transformation product may take only a few milliseconds to grow to its full size, before impingement occurs. Such rapid rates can be measured only by special experimental techniques such as pulse heating. In this technique, a specimen that is quenched to a low temperature to retain the high temperature phase in a metastable condition, is reheated for a few milliseconds by means of a pulse of energy generated from a capacitor-discharge or a laser beam. This technique allows the study of the massive transformation kinetics at a constant temperature, in contrast to the experiments of continuous cooling at different rates outlined earlier.

TABLE 9.1
Approximate Growth Rates for Different Transformations

	Type of transformation	*Growth rate, mm s^{-1}*
1	Growth controlled by long range diffusion of substitutional solute	10^{-3}
2	Growth controlled by long range diffusion of interstitial solute	$10^{-3} - 10^{-2}$
3	Polymorphic transformations and recrystallization	$10^{-3} - 10^{-1}$
4	Massive transformations	10
5	Martensitic transformations	10^{6}

9.2 THE DRIVING FORCE FOR THE MASSIVE TRANSFORMATION

The free energy-composition curves for the α and β phases of a binary system are schematically shown in Fig. 9.3. The common tangent xy to the two curves delineates the composition ranges corresponding to α, $\alpha + \beta$ and β phases. Segment ba falling in the α region typically represents the free energy change for the $\beta \rightarrow \alpha$ transformation without a change in composition $\bar{c}$. If the composition of the alloy were to lie in the two phase region, for example, c' in Fig. 9.3, a finite driving force $b'a'$ still exists for the $\beta \rightarrow \alpha$ transformation without a composition change, even though the transformation of β to two phases of composition of $c_{\alpha\beta}$ and $c_{\beta\alpha}$ has a larger driving force $b'a''$. The driving force for the transformation without a compositional change decreases, as the overall composition shifts to the right in Fig. 9.3, until it becomes zero at the point of intersection of the two free energy

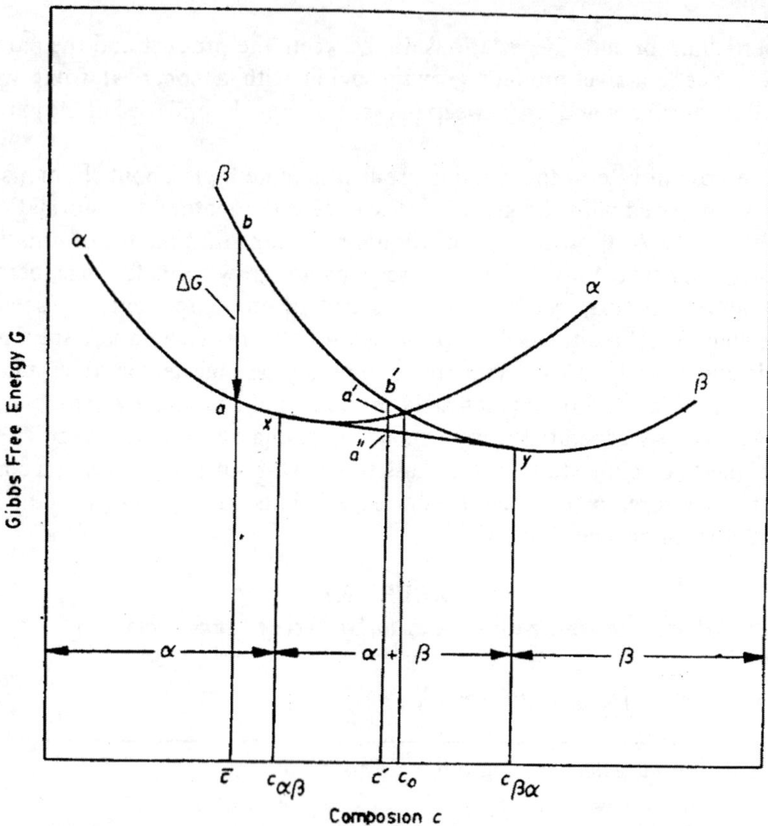

Fig. 9.3 Free energy of α and β phases as a function of composition in a binary system.

curves at composition c_0. To a first approximation, c_0 lies midway between $c_{\alpha\beta}$ and $c_{\beta\alpha}$.

The locus of c_0 as a function of temperature is shown for Cu–Zn alloys in the phase diagram in Fig. 9.4. T_0 is the temperature at which the driving force for the $\beta \rightarrow \alpha$ transformation in an alloy of composition c_0 is zero. For $c_0 = 38$ at.% Zn, $T_0 = 1000$ K and the free energy change for the $\beta \rightarrow \alpha$ transformation as a function of temperature is given by

$$\Delta G = -1670 + 1.67T \text{ J mol}^{-1} \tag{9.1}$$

9.3 THE MECHANISM AND THE KINETICS OF MASSIVE GROWTH

Growth Rates from Pulse Heating Experiments

Measurements of the isothermal growth rates of the massive transformation in a Cu-38 at.% Zn alloy have been carried out by Karlyn, Cahn and Cohen. The alloy is solutionized in the single phase β region and quenched to room temperature. Some α forms during the quench, but most of the β is retained in the metastable condition. The alloy is then pulse heated to temperatures ranging from 380 to 510°C. This temperature range corresponds to

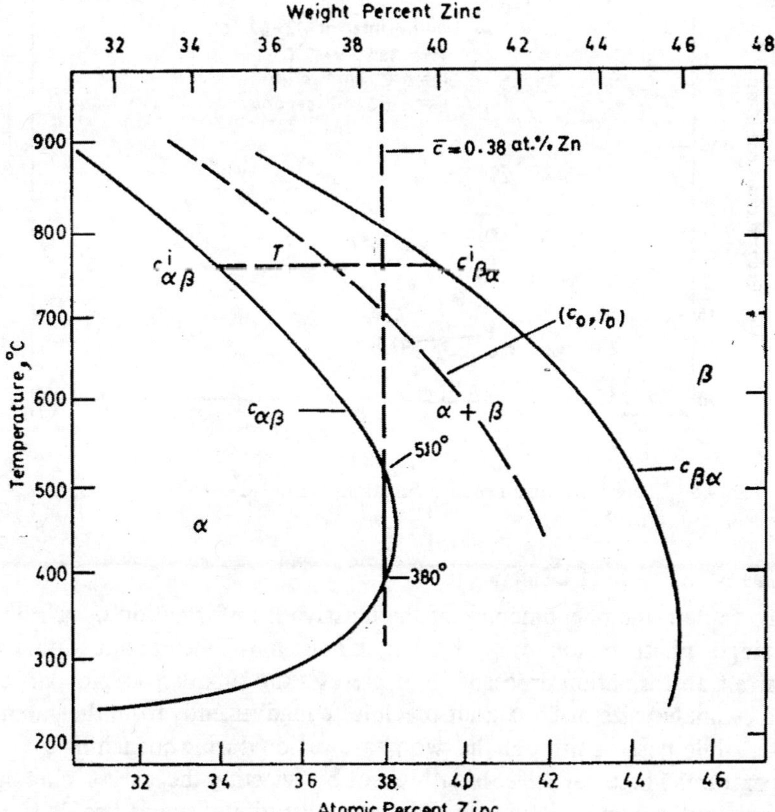

Fig. 9.4 The phase diagram of the Cu–Zn system, depicting the locus of the composition c_0, at which the free energies of α and β phases are equal.

the single phase α region for this composition, see the phase diagram in Fig. 9.4. The massive reaction $\beta \rightarrow \alpha_m$ takes place. A number of specimens are held at each reheating temperature for different lengths of time (in the range of milliseconds) and then quenched back to room temperature. The largest dimension of the largest patch of the massive product is measured from each sample. This is done up to a fraction-transformed $X = 0.3$ only, to minimize the effect of impingement of neighbouring patches.

A typical plot of the growth distance versus time of holding at ~ 417°C is shown in Fig. 9.5. The linear relationship indicates a constant growth rate of 8 mm s^{-1}, after an initial delay time τ of 5 ms. The growth rate increases monotonically on increasing the temperature from 380° to 510°C.

On pulse heating a sample above 510°C into the two-phase ($\alpha + \beta$) region, the massive reaction $\beta \rightarrow \alpha_m$ does not occur. This is somewhat surprising, as there is no discontinuous change in the driving force, as we cross into the two-phase region to the right of the composition $c_{\alpha\beta}$ in Fig. 9.3.

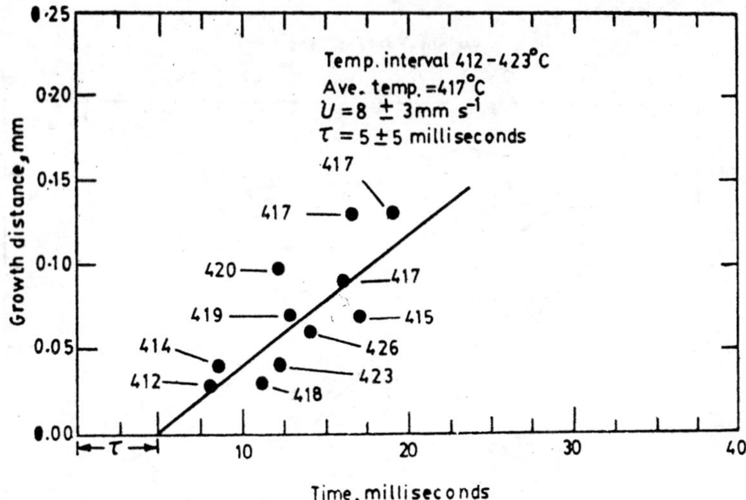

Fig. 9.5 The growth distance as a function of time at 417°C during massive growth.

The Mechanism of Growth

To explain the phenomenon of the massive transformation occurring in the single phase region only, Karlyn, Cahn and Cohen assumed that the massive transformation proceeds from pre-existing submicroscopic particles of an estimated size of 80 Å, that precipitate inadvertently from the parent β phase, while passing through the two-phase region during quenching to room temperature. These particles initially form by rejecting the excess zinc into the adjacent α matrix, giving rise to a concentration-distance profile shown schematically in Fig. 9.6(a). l is the effective thickness of the diffusion zone surrounding the spherical precipitate particle of radius r_i. $\bar{c}$ is the initial matrix composition before the precipitate particles form. $\overset{i}{c}_{\alpha\beta}$ and $\overset{i}{c}_{\beta\alpha}$ are the initial particle and surrounding matrix compositions corresponding to some temperature T in the two phase region, as indicated in Fig. 9.4.

On reheating into the α or $(\alpha + \beta)$ region, these particles continue to grow. If the reheating takes the alloy into the α region, $c_{\alpha\beta} > \bar{c}$ and further growth of the particle to radius r changes the concentration-distance profile as shown in Fig. 9.6(b). If the reheating is into the $(\alpha + \beta)$ region, $c_{\alpha\beta} < \bar{c}$ and the profile is modified as shown in Fig. 9.6(c).

The overall mass balance during growth on reheating is given by

$$\frac{4}{3}\pi r_i^3(\bar{c} - \overset{i}{c}_{\alpha\beta}) + \frac{4}{3}\pi(r^3 - r_i^3)(\bar{c} - c_{\alpha\beta}) + 4\pi r^2 l(\bar{c} - c_{\beta\alpha})/2 = 0 \tag{9.2}$$

The above is valid for $l \ll r$. Using $y = r/r_i$ and substituting for r in terms of r_i in Eq. 9.2, we obtain an expression for l:

$$l = \frac{2r_i}{3}\frac{(c_{\alpha\beta} - \overset{i}{c}_{\alpha\beta}) + y^3(\bar{c} - c_{\alpha\beta})}{y^2(c_{\beta\alpha} - \bar{c})} \tag{9.3}$$

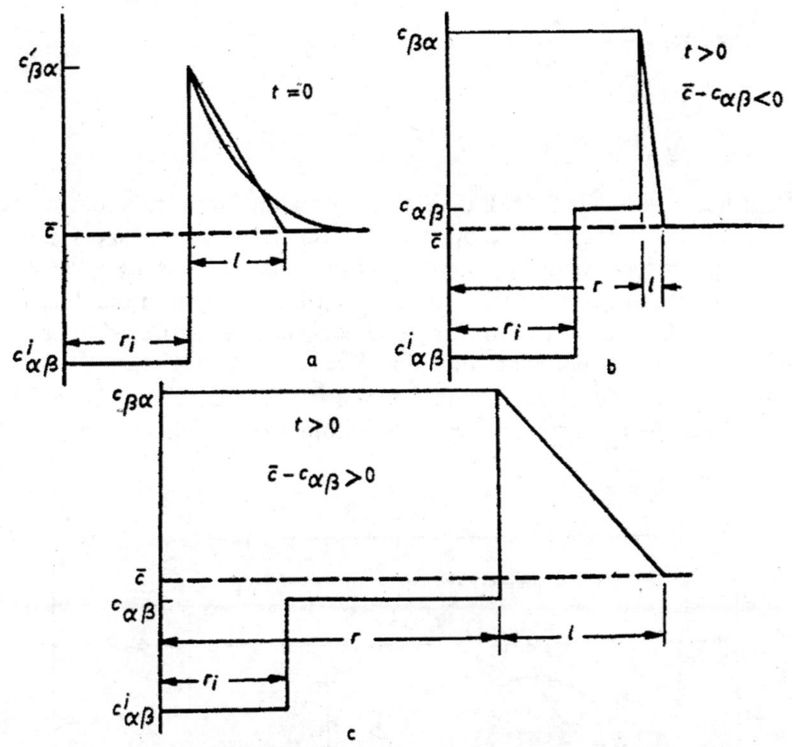

Fig. 9.6 Concentration-distance profile around a particle precipitated during quenching: (a) before reheating, (b) after reheating into the α region, and (c) after reheating into the $(\alpha + \beta)$ region.

When the reaction temperature lies in the α region, $(\bar{c} - c_{\alpha\beta}) < 0$. Therefore, l decreases with growth which is diffusion controlled (see Fig. 9.6(b)) and the growth rate increases. The excess zinc build up in the matrix is wiped out eventually and the growth rate attains the value characteristic of interface-controlled growth. The transformation then becomes *truly massive*. The delay time observed in Fig. 9.5 is identified with the time needed to wipe out the excess zinc build up.

When the reaction temperature is in the $(\alpha + \beta)$ region, $(\bar{c} - c_{\alpha\beta}) > 0$. Then l increases with particle growth (Fig. 9.6(c)) and the growth rate decreases with time. The diffusion-controlled growth continues and the massive transformation cannot get started in the two-phase region. Thus the model explains the observed experimental behaviour in Cu–Zn alloys.

The Temperature Dependence of the Growth Rate

As the massive transformation proceeds by interface-controlled growth, Eq. 5.2 (see page 73) is applicable. The supercooling ΔT is large ($\sim$200-300°C) for massive transformations in Cu–Zn alloys. So, we can use the approximation $v \mid \Delta g \mid \gg kT$ in Eq. 5.2 and obtain the simplified form given in Eq. 5.9:

$$U = \lambda_j \nu \exp\left(-\frac{\Delta G_D}{kT}\right)$$

$$= \frac{D_b}{\lambda_j}.$$

A plot of log U versus $1/T$ yields the activation energy for the rate controlling step, which is the transfer of atoms across the interface by short range diffusion. The experimentally determined value of ΔG_D is 61 kJ mol^{-1}, which is significantly lower than that for lattice diffusion in the β phase: 79.4 kJ mol^{-1}. As pointed out in Chapter 5, ΔG_D can be considered to be the free energy barrier for boundary diffusion.

A plot of U as a function of temperature using Eq. 5.2 is shown in Fig 9.7. The maximum in the growth rate falls within the two-phase region. As

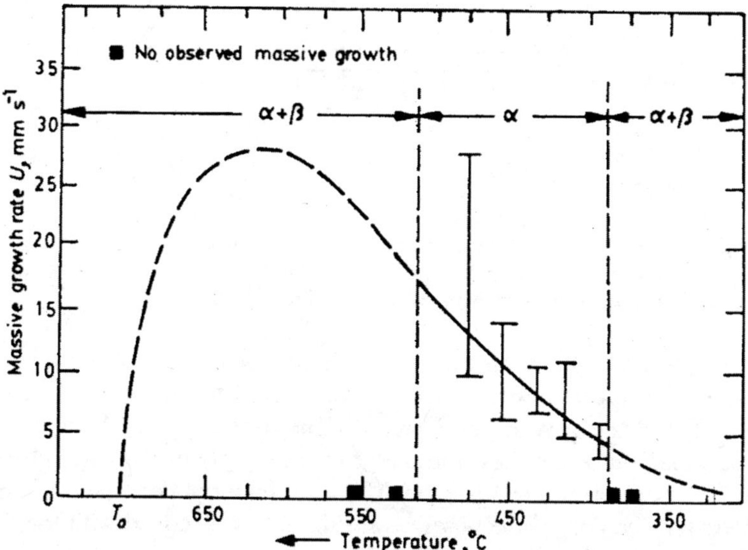

Fig. 9.7 The massive growth rate as a function of temperature calculated from Eq. 5.2. The experimental points fall in the region where $\nu \mid \Delta g \mid \gg kT$.

discussed above, the massive reaction does not get started in the two-phase region from the inadvertantly-precipitated α particles. The plot in Fig. 9.7 implies that the transformation mechanism is the same for a massive transformation, as it is for a non-martensitic polymorphic transformation in a pure substance. Uncoordinated transfer of atoms across an incoherent interface characterizes both these transformations. The faster growth rates observed in massive transformations are then attributable to the larger driving force available for such transformations.

FURTHER READING

D.A. Karlyn, J.W. Cahn and M. Cohen, *Trans. Met. Soc. AIME*, **245**, 197 (1969).

Symposium Papers on Massive Transformations, *Metall. Trans. A*, **15A**, 411-447 (1984).

EXERCISE

9.1 Compare the types of phase changes in Buerger's classification with the massive transformation and comment.

10

Recovery, Recrystallization and Grain Growth

Deformation and annealing are among the most important processing methods adopted for engineering metals and alloys to effect structural and corresponding property changes. The processes of recovery, recrystallization and grain growth refer to the microstructural changes that take place on heating a cold worked material.

In crystalline materials, the density of point imperfections (vacancies and interstitialcies) and dislocations increases with the extent of plastic deformation at temperatures below about $0.3–0.4T_m$, where T_m is the melting point of the material in kelvin. All such structural imperfections have strain energy associated with them. So, the plastically deformed state has a higher free energy than the undeformed (or annealed) state. Between 1 and 15% of the work of deformation remains stored in the material in the form of these imperfections. The amount of *stored energy* is influenced by the extent, temperature, rate and type of deformation given and also by the composition, purity and grain size of the material. This stored energy provides *the driving force* for various relaxation processes that occur during subsequent annealing.

Recovery is the annihilation and the rearrangement of imperfections within the deformed material without the movement or migration of high angle boundaries. *Recrystallization* is the nucleation and growth of new, strain-free grains by the migration of high angle boundaries. In *grain growth*, the grain boundaries already present simply migrate in such a direction as to increase the average grain size of the material. In none of these three phenomena, a change in composition or crystal structure is involved, as would be the case in a typical phase transformation. However, the boundary migration processes have much in common with similar growth processes in phase transformations.

The essential characteristics of recovery, recrystallization and grain growth are discussed in the ensuing three sections.

10.1 RECOVERY

There are two basic processes taking place during recovery:

1 Annihilation of point defects and formation of special configuration of point defects, and

2 Annihilation and rearrangement of dislocations.

As the annealing temperature is increased, the first of the two processes sets in and the second follows with a further increase in temperature. However, it should be noted that, depending on the temperature of deformation, a part of the recovery processes may occur during the deformation itself. This phenomenon is known as *dynamic recovery*. The structural changes during recovery occur more or less simultaneously throughout the deformed matrix.

Annihilation of Point Imperfections

Here, the excess point imperfections produced by plastic deformation anneal out in various ways. A pair of a vacancy and an interstitialcy can annihilate each other. The point imperfections also find a sink at high angle (incoherent) boundaries and at edge dislocations. Excess vacancies disappear at edge dislocations, making them to climb up. Interstitialcies disappear in a similar manner at edge dislocations, causing the dislocation line to climb down.

This part of the recovery process can be followed experimentally by measurements of the electrical resistivity of the sample as a function of time at the annealing temperature. As the free electrons in metals are scattered by point imperfections, their annealing out results in a decrease in the electrical resistivity of the material. The changes in electrical resistance of copper cold worked at 4.2 K and subsequently annealed isothermally at various higher temperatures is shown in Fig. 10.1. The resistance decreases sharply in the early stages and then more gradually with increasing annealing time, approaching a constant value characteristic of the temperature of annealing.

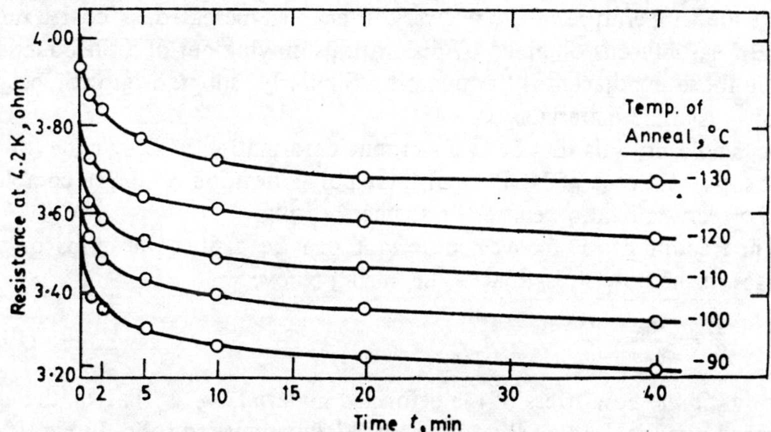

Fig. 10.1 The decrease in the electrical resistance of a cold-worked copper specimen during recovery.

Dislocation Rearrangements

As the annealing temperature is increased, the recovery process becomes more of annihilation and rearrangement of dislocations. A pair of a positive and a negative edge (or screw) dislocation with the same magnitude of the Burgers vector can annihilate each other. An edge-type pair lying on the same slip plane can come together by gliding and annihilate each other. If the edge dislocations are lying on adjacent parallel planes, their coming together may result in a row of vacancies or interstitialcies. These point imperfections can be annihilated by a further reaction with other point imperfections. In the case of a screw dislocation pair, mutual annihilation is possible, even if they are lying on different slip planes. Cross-slip can bring together the screw dislocations to the same slip plane to be mutually annihilated.

After annihilation of dislocations of opposite sign, excess dislocations of the same sign are left over in the crystal. The elastic strain energies of such dislocations can be reduced further, when they arrange themselves in low angle boundaries. Excess edge dislocations form a tilt boundary, whereas excess screw dislocations give rise to twist boundaries. This process is known as *polygonization*. When a single crystal is deformed by bending in one direction, excess edge dislocations of the same sign are produced on parallel slip planes. On annealing, the edge dislocations form a number of tilt boundaries, converting the curved bent surface of the crystal to a number of planar surfaces tilted with respect to one another.

In polycrystalline metals, during the third stage of deformation, dynamic recovery is manifest in the *formation of cells or subgrains* within the grains. The walls of the cells are relatively broad and have a high dislocation density, while the interior is nearly free of dislocations. During the recovery anneal, in the interior of the cells, the dislocation density decreases further by dipole annihilation and dislocation movement into the cell walls, which become more sharply delineated.

As the temperature of the recovery anneal is increased, a coarse polygonized structure is obtained by dislocations moving out of a tilt boundary to join those in adjacent tilt boundaries. Similarly, subgrain growth occurs in polycrystalline materials.

In single crystals that have undergone deformation by easy glide (or by basal slip in the case of HCP metals), the polygonization can be so complete that no recrystallization ensues on further heating.

The fraction of the recovery process X can be defined in terms of the changes in the flow stress that occur during recovery:

$$X = \frac{\sigma_m - \sigma}{\sigma_m - \sigma_0} \tag{10.1}$$

where σ_m is the flow stress of the deformed material, σ_0 is that of the undeformed material and σ is that of the partially recovered solid. In Fig. 10.2, $(1 - X)$ is plotted against time t at various isothermal annealing temperaratures for iron strained 5% in tension at 0°C. The kinetics are similar to that in Fig. 10.1.

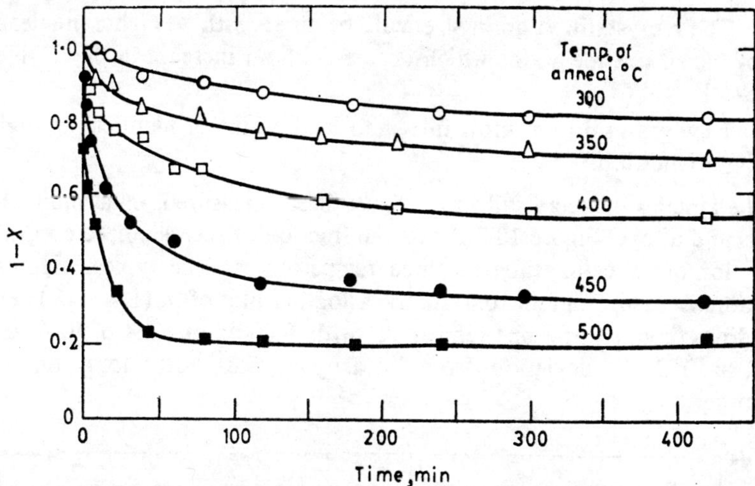

Fig. 10.2 The fraction of recovery X defined in terms of the flow strengths of iron, plotted as $(1 - X)$ versus time.

10.2 RECRYSTALLIZATION

Recrystallization can be described as the nucleation and growth of new, strain-free crystals by the progressive consumption of the cold-worked matrix. The growth occurs by the migration of high angle boundaries between the distorted matrix and the new crystals, until impingement occurs among the new crystals.

Experimental Characteristics

Some of the well-known "laws" of recrystallization are listed below:

1 A minimum deformation is necessary to cause recrystallization.

2 Recrystallization occurs at measurable rates at temperatures above 0.3–0.5T_m of the material, where T_m is the melting point in kelvin. The recrystallization temperature is arbitrarily defined as the temperature at which 50% of the cold worked matrix recrystallizes in 1 hr. It is strongly dependent on the purity of the metal. Very high purity metals recrystallize around 0.3T_m, whereas impure metals may recrystallize only around 0.5–0.6T_m.

3 As thermal activation is required for recrystallization, the time for recrystallization increases with decreasing annealing temperature.

4 With increasing degree of cold work, the stored energy increases and the recrystallization temperature decreases.

5 As the grain boundaries of the matrix play a role in the nucleation of strain-free crystals, finer initial grain sizes lower the recrystallization temperature.

6 A high deformation temperature raises the recrystallization temperature, as the stored energy decreases with increasing temperature of cold work.

7 The recrystallized grain size will be finer with a higher nucleation rate of the strain-free grains which occurs with an increase in the degree of cold work.

8 New grains do not grow into deformed grains of identical or slightly varying orientation.

The kinetics of recrystallization have been measured in a number of metals and alloys. Figure 10.3 shows the fraction of recrystallized copper as a function of annealing time at three temperatures. The typical sigmoidal variation is seen. Figure 10.4 shows a log-log plot of ln $(1/1 - X)$ versus annealing time t for zone-refined Al with 0.0068 at.% Cu 40% cold-rolled at 0°C. A deviation from linearity is seen after long times of annealing.

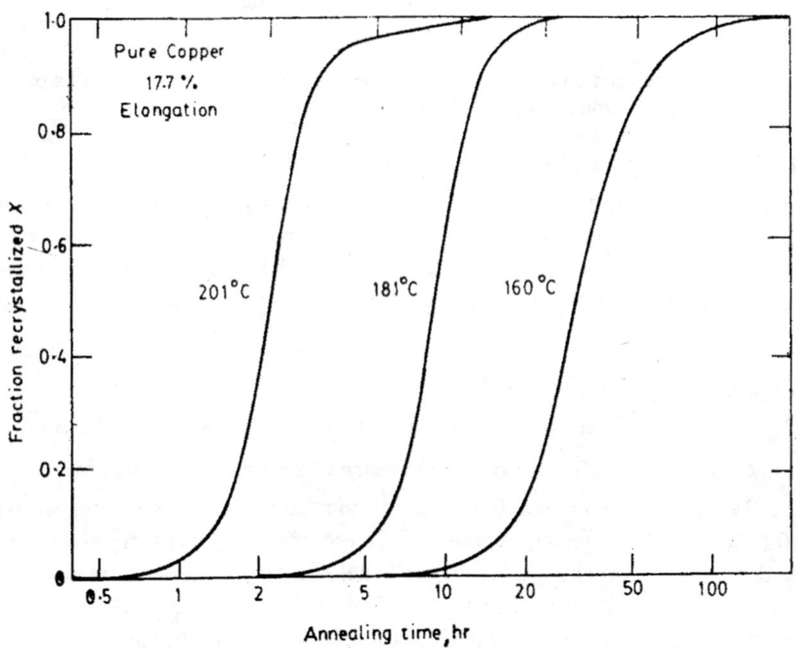

Fig. 10.3 The recrystallization kinetics in cold worked copper.

The nucleation and the growth rates during recrystallization have also been measured. The early work of Mehl and coworkers showed that the nucleation rate during recrystallization increases exponentially with time. On the other hand, the growth rate decreases with time, which has been attributed to the concurrent recovery taking place in deformed regions yet to recrystallize. Figure 10.5 shows that the reciprocal of growth rate varies linearly with annealing time in a titanium alloy.

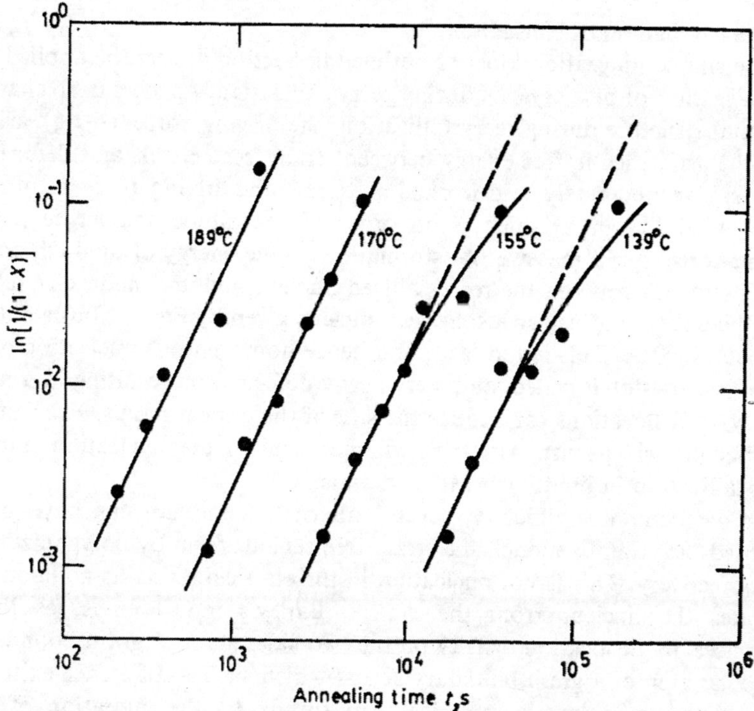

Fig. 10.4 A log-log plot of ln (1/1 — X) versus annealing time for cold rolled aluminium.

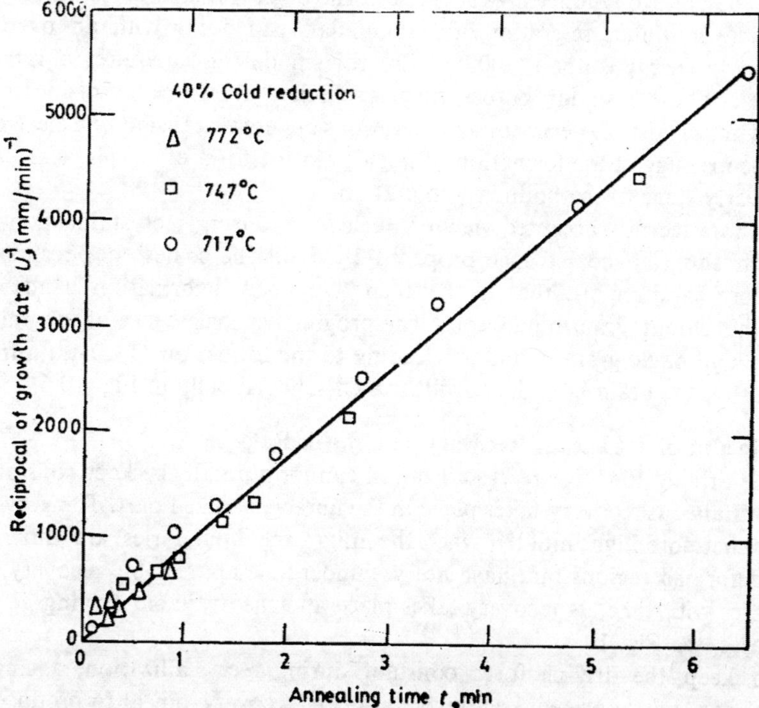

Fig. 10.5 The reciprocal of the growth rate during recrystallization of a titanium alloy varies linearly with annealing time.

The Mechanism of Nucleation

The simple nucleation kinetics outlined in Section 4.1 can be applied to the nucleation of new crystals during recrystallization. As there is no change in crystal structure during recrystallization, the driving force $(-\Delta g)$ arises from the difference in free energy between strain-free crystals and deformed crystals. For moderately cold-worked materials, the driving force is of the order of 10 MJ m^{-3}, which is an order of magnitude smaller than the driving force for a massive transformation. If the energy of the high angle grain boundary between the recrystallized nucleus and the matrix is 0.5 J m^{-2}, the critical radius for a spherical nucleus given by Eq. 4.3 turns out to be about 1000 Å. This is too large and hence homogeneous nucleation of a strain-free crystal is ruled out. Very heavy deformation resulting in a high density of dislocations can reduce the size of the critical radius. Even then, the difficulty will persist. Also, this will not explain the nucleation during recrystallization in lightly-worked materials.

To overcome this difficulty, several alternative mechanisms have been proposed. In Cahn's model, the strain-free regions form by polygonization during recovery. This is not nucleation in the classical sense, as no interface separates the nucleus from the matrix. Bailey's model envisages local differences in dislocation density on the two sides across a grain boundary. A circular area of a grain boundary across which such a difference exists is assumed to bulge into a spherical cap simply by the migration of the boundary into the region of higher dislocation density. As bulging takes place, the grain boundary area increases. The critical radius of curvature at which the continued migration of the boundary can occur with an overall decrease in energy is about 1000 Å. So, this model also requires a rather large area of the boundary across which a sufficient difference in dislocation density must exist. Experimental observations under the transmission electron microscope suggest the formation of nucleus in the form of a spherical cap at an early stage of the boundary migration.

A more recently-accepted view of nucleation during recrystallization is through subgrain coalescence proposed by Hu. The coalescence occurs in the micro-band regions that lie between two main deformation bands or near the parent grain boundaries. The progressive coalescence of subgrains near a high angle grain boundary, leading to the formation of a strain-free nucleus on the grain boundary is illustrated schematically in Fig. 10.6.

Effect of Simultaneous Recovery on Growth Rate

The driving force for recrystallization cannot normally be kept constant, as simultaneous recovery takes place in the unrecrystallized part. The growth rate is therefore high initially, when the migrating boundaries are moving into deformed regions that have not yet undergone appreciable recovery. It decreases with time, as recovery takes place and the available driving force decreases, see Fig. 10.5.

To keep the driving force constant during recrystallization, a single crystal can be deformed and given a prior recovery anneal to produce a

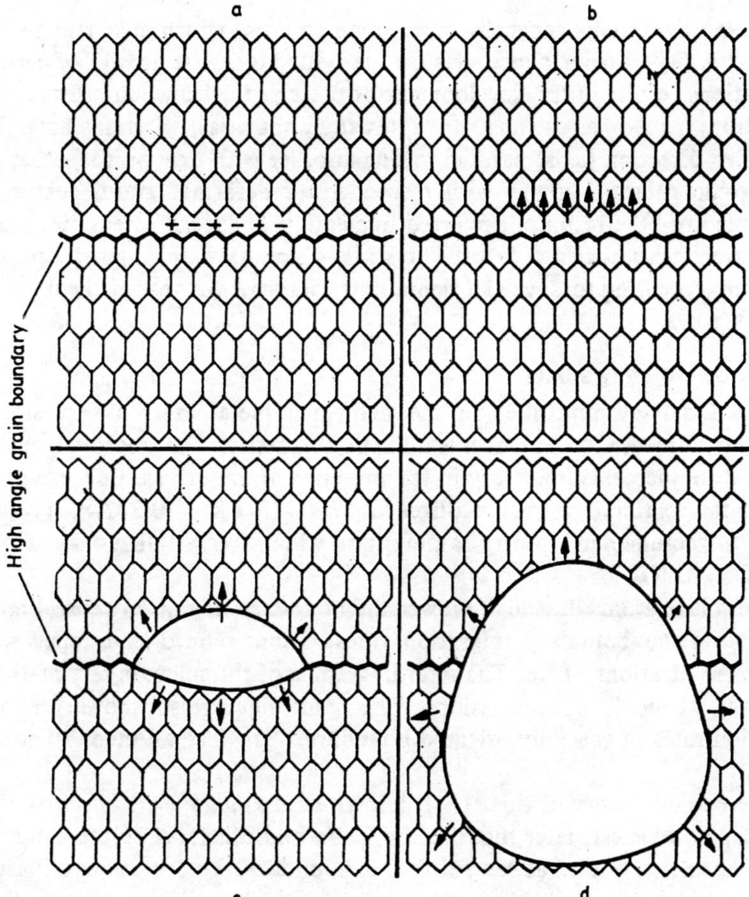

Fig. 10.6 Schematic illustration of the formation of a recrystallized nucleus by subgrain coalescence.

stable polygonized structure. Further deformation in one corner of the crystal can initiate the recrystallization. The boundary then migrates into the already polygonized region at a constant rate under conditions of constant driving force.

Effect of Orientation on Boundary Migration

The migration rate is found to be a function of the orientation difference between the two grains separated by the boundary. Certain specific orientations give rise to high migration rates. The favoured orientations are:

(a) *FCC crystals*

1 A rotation of one of the crystals by 38° and 22° about a ⟨ 111 ⟩ axis common to the two crystals

2 a rotation of 28° about a ⟨ 100 ⟩ axis.

(b) *BCC crystals*

A rotation of 30° about a ⟨ 110 ⟩ axis.

These orientations generally correspond to those which give rise to the highest density of coincidence sites on the boundary. Fast growth of certain orientations results in the development of preferred orientation after recrystallization, known as *texture.* In a textured material, all grains have this low index direction (axis) parallel to one another with one or the other of the specific rotation angles with respect to the deformed grain. Texture is not necessarily the result of preferred nucleation of special orientations. All orientations may nucleate, but the crystals, which have the highest growth rates corresponding to special orientations, outgrow the rest of them giving rise to texture.

The Solute Drag Effect

The dissolved impurities in the deformed metal exert a very strong retarding influence on the recrystallization kinetics. The recrystallization temperature increases markedly in the presence of impurities. For example, very pure aluminium recrystallizes at 75°C (348 K $= 0.37T_m$), as compared to commercial purity aluminium which recrystallizes at 275°C (548 K $= 0.59T_m$).

The solute atoms drastically lower the boundary migration rates. Figure 10.7 shows the boundary migration rates of zone refined lead doped with small concentrations of tin. The retarding effect of the solute is very marked, especially at small concentrations. The same figure also shows that the migration rates of specially oriented boundaries are less affected by solute atoms.

The earliest theory of this effect, known as the *solute drag effect*, is that of Lucke and Detert, later modified by Cahn and others. A solute atom has a binding energy V that attracts it to the boundary. Under an equilibrium distribution, the concentration of the solute at the boundary c_b is larger than the average concentration c_0:

$$c_b = c_0 \exp\left(\frac{V}{kT}\right) \tag{10.2}$$

When the boundary migrates, the solute atoms are also dragged behind it due to the attraction. A steady state velocity of the boundary is attained when the force due to the free energy difference between the strain-free and the deformed crystals pulling forward is equal to the dragging force that is pulling back the boundary. The velocity U of the boundary under such conditions is given by

$$U = \frac{\Delta g\, D}{kT\Gamma} \tag{10.3}$$

where

Δg is the driving force per unit volume,
D is the bulk diffusion coefficient of the solute, and
Γ is the number of solute atoms per unit area of the boundary.

Γ is given by

$$\Gamma = \alpha c_0 \exp\left(\frac{V}{kT}\right) \tag{10.4}$$

where α is a geometrical factor. Substituting Eq. 10.4 in 10.3, we have

$$U = \frac{\Delta g\, D}{kT \alpha c_0} \exp\left(-\frac{V}{kT}\right) \tag{10.5}$$

This result indicates that the steady state velocity is proportional to the driving force and the diffusion coefficient of the solute, but inversely proportional to the concentration of the solute. This explains the drastic decrease in the growth rate even for small concentrations of the solute, as shown in Fig. 10.7.

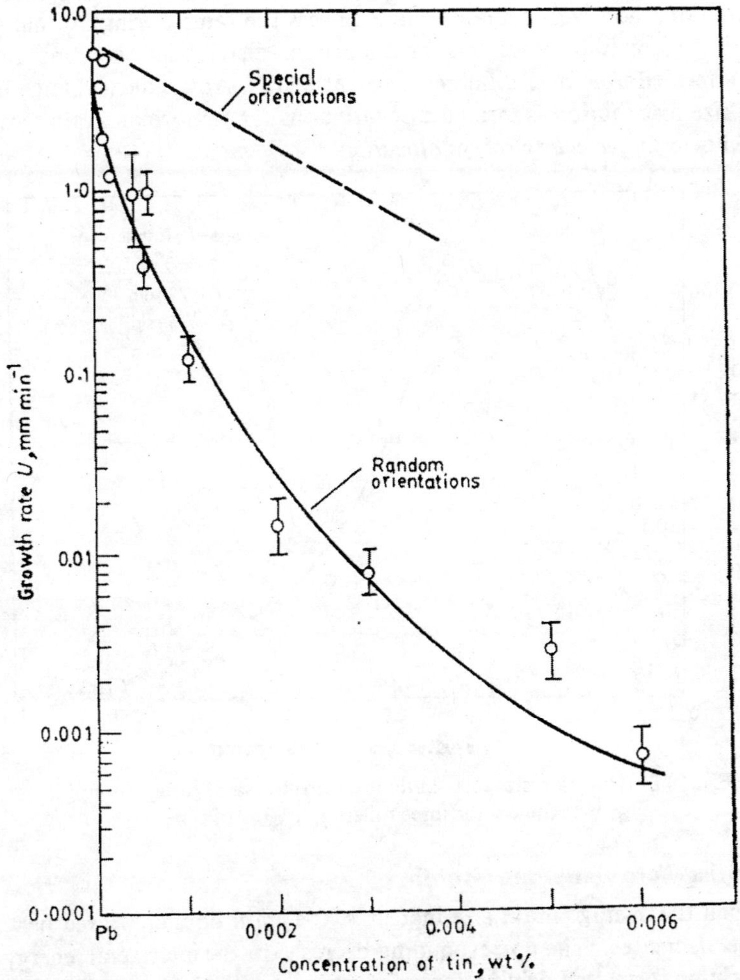

Fig. 10.7 Growth rates during recrystallization of zone refined lead doped with tin.

The force f arising from the interaction between the solute atoms and the boundary has a maximum value:

$$f \sim \frac{V}{2r} \tag{10.6}$$

where r is the atomic radius. When the forward force arising from $|\Delta g|$ exceeds f, *breakaway* of the boundary from the solute atmosphere occurs. The velocity of the boundary then corresponds to the interface-controlled growth rate. It is easily seen that breakaway occurs at large driving forces or at small concentrations of solute atoms.

10.3 GRAIN GROWTH

Grain growth refers to the increase in the average grain size on further annealing, after the recrystallization process is complete. The grain size distribution, however, remains more or less the same during normal grain growth. Figure 10.8 shows this invariance in distribution in zone refined iron annealed for three different times at 650°C. Any radical change in the grain size distribution is termed discontinuous or abnormal grain growth, also known as *secondary recrystallization*.

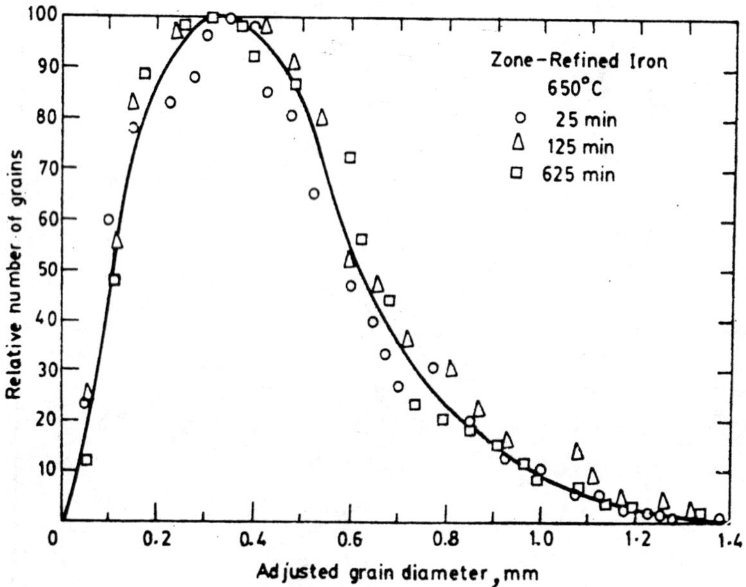

Fig. 10.8 The grain size distribution remains the same, after the grain growth anneal for three different lengths of time.

Driving Force for Grain Growth

When the average grain size increases, the grain boundary area per unit volume decreases. The corresponding decrease in the interfacial energy per unit volume forms the driving force for grain growth. This driving force is about an order of magnitude smaller than that for recrystallization.

Consider the idealized case of a spherical grain of radius R. Here, the grain boundary energy per unit volume is $4\pi R^2\sigma \Big/ \frac{4}{3}\pi R^3 = 3\sigma/R$. For a curved boundary, the radius of curvature is positive on one side and negative on the other. The free energy difference per unit volume is

$$|\Delta g| = C\sigma\left(\frac{1}{R_1} - \frac{1}{R_2}\right) \tag{10.7}$$

where C is a constant. If $R_1 = R_2 = R$

$$|\Delta g| = \frac{2C\sigma}{R} \tag{10.8}$$

The Parabolic Law of Grain Growth

The grain growth kinetics can be described by the following empirical equation (when the initial grain size is small enough to be neglected):

$$\bar{D} = C_1 t^n \tag{10.9}$$

where $\bar{D}$ is the average grain diameter and C_1 is a constant.

A simple growth model described below shows that the time exponent n should be $\frac{1}{2}$. If we assume that the rate of change of $\bar{D}$ is proportional to Δg, then

$$\frac{d\bar{D}}{dt} = C_2 |\Delta g| \tag{10.10}$$

From Eqs. 10.10 and 10.8, we can write

$$\frac{d\bar{D}}{dt} = \frac{C_3\sigma}{\bar{D}} \tag{10.11}$$

Integrating Eq. 10.11,

$$\int_{\bar{D}_0}^{\bar{D}_t} \bar{D}\, d\bar{D} = C_3\sigma \int_0^t dt$$

or

$$(\bar{D}_t^2 - \bar{D}_0^2) = 2C_3\sigma t \tag{10.12}$$

where the subscripts t and 0 refer to time. At small initial grain sizes, $\bar{D}_t \propto t^{1/2}$. This parabolic growth law has been observed experimentally. However, the time exponent is often less than 0.5, especially at low temperatures of annealing.

Pinning by Second Phase Particles

Grain growth can be hampered in the presence of inclusions, voids and second phase particles. Figure 10.9 shows schematically a grain boundary before it meets a precipitate particle and after. In positions 1 and 3, the

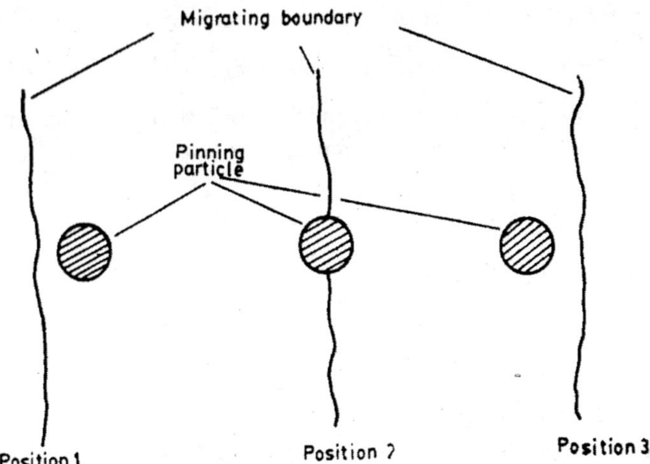

Fig. 10.9 The pinning action of a precipitate particle on a migrating grain boundary.

combined interfacial energy of the grain boundary and the precipitate particle is the same, before the boundary meets the particle (position 1) and after it has left the particle behind (position 3). In position 2, however, when the particle lies in the boundary, the grain boundary area is reduced by an amount equal to the cross-sectional area of the particle. This reduction in energy provides an attractive force between the particle and the boundary. When the boundary is to move away from position 2 to position 3, the energy of the boundary has to increase again to the initial value corresponding to position 1. Such an increase in energy is resisted and manifests itself as the *pinning action* of the precipitate particle on the boundary.

With N_v precipitate particles of average radius r_p per unit volume, the driving force in Eq. 10.8 is reduced to

$$| \Delta g | = 2C\sigma \left(\frac{1}{R} - N_v \pi r_p^2 \right) \tag{10.13}$$

The volume fraction of the particles X is

$$X = \frac{4}{3} \pi r_p^3 N_v \tag{10.14}$$

Substituting for N_v from Eq. 10.14 in 10.13, we obtain

$$| \Delta g | = 2C\sigma \left(\frac{1}{R} - \frac{3X}{4r_p} \right) \tag{10.15}$$

If $| \Delta g |$ is positive, grain boundary migration will continue although at a reduced rate. If $| \Delta g | \leqslant 0$, the migration rate is zero and the grain growth will cease at a finite grain size. This occurs when

$$R \geqslant \frac{4r_p}{3X} \tag{10.16}$$

Note that the critical stage in pinning occurs at small grain sizes, when the volume fraction of the precipitate particles is large or the average particle size is small.

It may also be noted that the boundary migration process is basically the same during recrystallization and grain growth. The solute drag effect discussed under recrystallization and the pinning by particles discussed here are applicable to both recrystallization and grain growth.

If the second phase particles go into solution on heating the material, grain growth often starts again. This indicates that the solute drag effect on the migrating boundaries is here less effective than the pinning effect of the precipitate particles.

FURTHER READING

H. Hu, *Metallurgical Treatises*, The Metallurgical Society of AIME, Warrendale, PA, p. 385 (1981).

K. Lucke and K. Detert, *Acta Metall.*, **5**, 628 (1957).

EXERCISE

10.1 Hard-rolled pure copper recrystallizes by 50% on heating for 3 min at 150°C or 80 min at 100°C. What is the time for 50% recrystallization at 50°C?

11

Martensitic Transformations

The hard, brittle phase produced in steels on quenching was named martensite, after the German metallurgist A. Martens. In recent decades, the term has come to include the products obtained as a result of a special type of solid-state transformation with certain unique characteristics. The question of which of these characteristics are essential in the definition of a martensitic transformation has undergone changes from time to time, as more and more transformations with similar characteristics have come to light in ferrous and nonferrous alloys, as well as in nonmetallic crystals.

Recent attempts to classify the known transformations and define the term martensite are described, after introducing the general characteristics of the transformation. The essential morphological and kinetic characteristics are discussed in some detail, with particular reference to iron base alloys. The model that has been developed to compute the isothermal transformation curves in iron alloys is described. The phenomenological theory of the crystallography of martensitic transformations is briefly reviewed. The special features of nucleation and growth, including the Olson-Cohen mechanism of nucleation, are described in the end.

11.1 GENERAL CHARACTERISTICS

Diffusionless Nature

During a martensitic transformation, each atom tends to retain its original neighbours. There is no interchange of sites between atoms as would be the case in a typical diffusion process. This rules out both interface-controlled growth by short range diffusion and diffusion-controlled growth by long range diffusion. In some martensitic transformations, however, there may be *incidental diffusion*, such as in the simultaneous tempering of the martensite formed in steels. In bainitic transformations, long range diffusion is *required* for the transformation to proceed. This latter category then clearly does not satisfy the criterion of being diffusionless in nature.

The lack of long range diffusion, of course, means that there is no change in composition during martensitic transformations. The lack of short range diffusion and the fact that each atom tends to retain its neighbours implies that an ordered phase will transform martensitically to another ordered phase.

Displacive Nature

The martensitic transformation is displacive in nature and occurs by a process of shear. The shear causes surface upheavals, when the transformation product extends to the free surface of the material. Fiducial lines or scratches on a prepolished surface tilt out of position after the transformation, without losing continuity across the interface as illustrated in Fig. 4.8. The interface is a well-defined crystallographic plane, with characteristic Miller indices. It is known as the *habit plane*.

The Habit Plane

Observations such as the continuity of the fiducial lines across the interface have led to the concept that the habit plane is an undistorted and unrotated plane. This description will be valid, if the transformation strain can be factorized into a *simple shear parallel to the habit plane and a dilatation perpendicular to it*, as illustrated in Fig. 11.1(b) and (c). In addition, *atom shuffles* can take within the unit cell, as indicated in Fig. 11.1(d). Figure 11.1(e) brings together the shear, the dilatation and the atom shuffle. All these operations do not distort the habit plane. As will be seen in detail later, if the transformation strain cannot be factorized into these, a second shear not parallel to the habit plane may also be visualized, provided that it takes place in such a manner as not to distort the habit plane *in a macroscopic sense*.

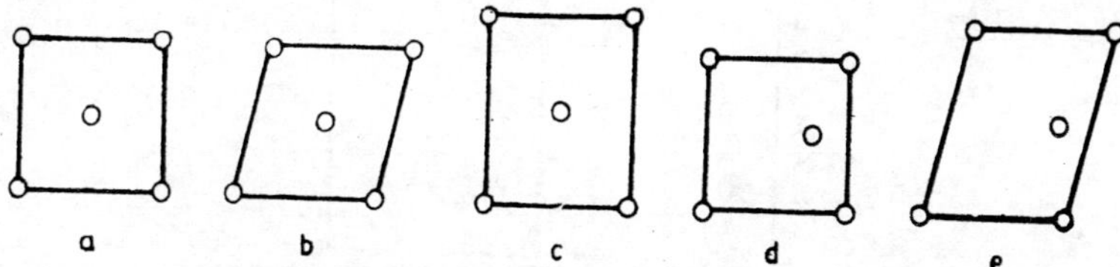

Fig. 11.1 The habit plane is the base of the cube. (a) Undeformed lattice, (b) shear parallel to the habit plane, (c) dilatation perpendicular to the habit plane, (d) atom shuffle within the unit cell, and (e) combination of (b), (c) and (d).

Strain Energy due to Shape Change

When the transforming region changes shape within the parent phase, the shear strains produced result in elastic strain energy. The shear strain in steels is quite appreciable, of the order of ~0.2. The accompanying large strain energy has a pronounced inhibiting effect on the transformation. Also, the balance between the strain energy and the surface energy results in a plate-like or a lath-like particle.

Kinetic Characteristics

Martensite often forms athermally, i.e., without any apparent assistance from thermal energy. The transformation starts at a well defined temperature M_s and is virtually complete at a lower temperature M_f, Fig. 11.2(a). In the absence of interfering reactions, the reverse transformation likewise starts at a well-defined temperature A_s and is complete at A_f, Fig. 11.3.* The magnitude of the thermal hysteresis depends on the system, generally being large for systems where the strain energy of the transformation is large. Thus, the hysteresis is large for Fe–Ni alloys and small for Au–Cd alloys, as illustrated in Fig. 11.3. The transformation can be induced above M_s, if the parent phase is plastically deformed. The highest temperature up to which plastic deformation of the parent phase can induce the transformation is termed M_d, with a corresponding definition for A_d.

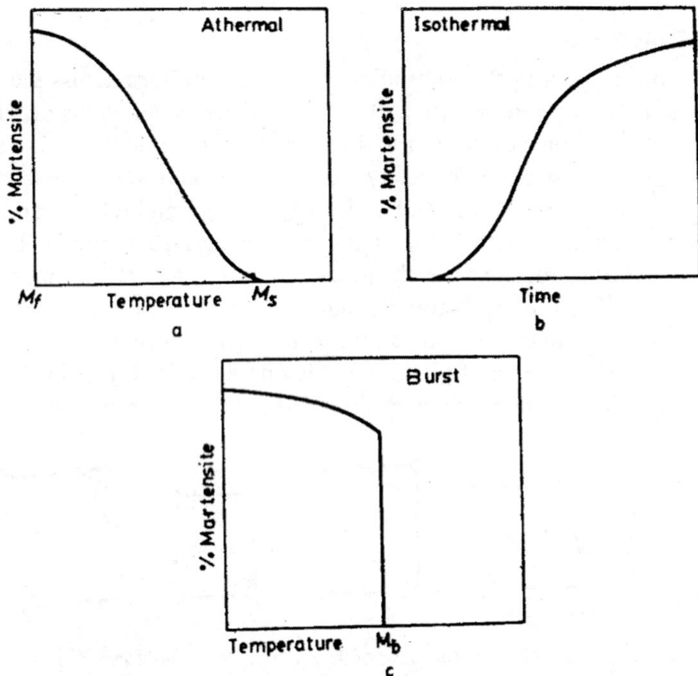

Fig. 11.2 The three kinetic modes: (a) athermal, (b) isothermal, and (c) burst.

In some alloys, martensitic transformation occurs isothermally, i.e., at a constant temperature as a function of time, Fig. 11.2(b). Here, thermal activation is clearly necessary for the nucleation of the martensitic phase.

In some other alloys, especially when the transformation range is below room temperature, the martensitic transformation starts abruptly in the form of a burst at a characteristic temperature M_b, Fig. 11.2(c). The amount of martensite that forms at M_b may be between 10 and 60%.

*A stands for austenite, the parent phase in steels. In many places in this chapter, we refer to the parent phase as austenite.

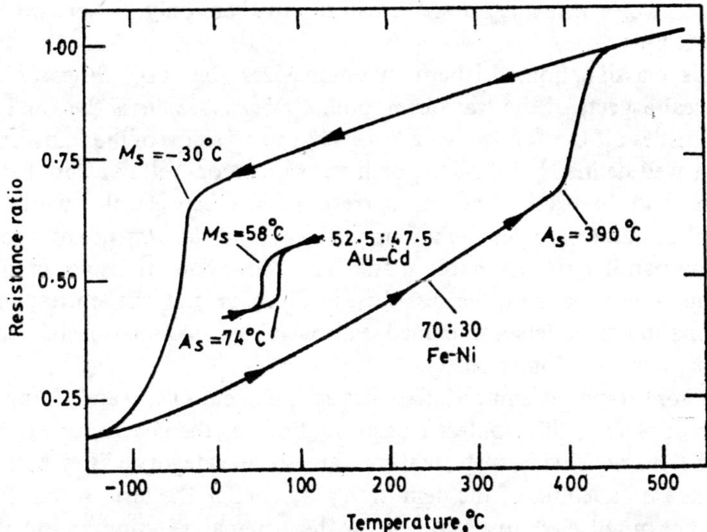

Fig. 11.3 The temperature hysteresis between the martensitic transformation and the reverse transformation in Fe—Ni and Au—Cd alloys. (The electrical resistivity of martensite is lower than that of the parent phase.)

Nucleation and Growth Processes

Heterophase fluctuations, which play a prominent role in the classical nucleation theory, are not involved in martensitic nucleation. Here, the kinetics of nucleation are governed by the dynamics of the advancing boundary dislocations. Ordinarily, this motion is a high-velocity phenomenon that does not show a detectable time dependence. However, in isothermal nucleation of martensite, the observed time dependence is due to the fact that the stress required to move a dislocation has both an athermal and a thermal component. The latter can be assisted by thermal fluctuations, which results in time-dependent nucleation.

The growth rates of martensitic units have been measured using sophisticated electronic techniques. The rates are generally found to be very high, $\sim 10^6$ mm s^{-1}. The rates are independent of temperature and are of the same magnitude in different kinetic modes. These high rates preclude the possibility of growth being a rate controlling step in martensitic transformations.

Thermoelastic Behaviour

In some nonferrous alloys, martensite exhibits thermoelastic equilibrium, i.e., the plate attains a thickness characteristic of the temperature to which the alloy is cooled. If the alloy is cooled further, the plate thickens more; if the alloy is heated, the plate shrinks. Such alloys exhibit the *shape memory effect* and are currently studied for their potential use in various devices.

11.2 CLASSIFICATION OF MARTENSITIC TRANSFORMATIONS

Several classifications of martensitic transformations have been attempted in the last two or three decades. We will discuss here only one or two of the recent attempts.

In his classification, Lieberman emphasizes the crystallographic or geometrical aspects of the transformation. *Orthomartensite* is the ideal case, which satisfies all the following criteria: change of shape of the transforming region, a well-defined habit plane, definite orientation relationship between the parent and the product phases, correspondence between the two lattices, diffusionless nature, no change in composition, and the presence of a second shear non-parallel to the habit plane. The above conditions are gradually relaxed in describing the other less-than-ideal cases and the corresponding transformation products are named variously as: paramartensite, quasi-martensite and pseudomartensite.

In a more recent attempt at classification, Cohen, Olson and Clapp consider the displacive/diffusionless transformations as the general category, of which the martensitic transformations are a subcategory. They point out that undue broadening of the definition may change the term so much that a new name might have to be coined for the original reaction in steels responsible for hardening! We have already noted that there can be three components to a transformation strain that keeps the habit plane undistorted:

1 a shear parallel to the habit plane,
2 a dilatation perpendicular to the habit plane, and
3 an atom shuffle within the unit cell.

When shuffle displacements dominate, the displacive transformation is classified as a *shuffle transformation*. The kinetics of the transformation here is controlled by the shuffle displacements.

When the shear or the dilatation dominates, the transformation is termed as *lattice distortive*. This category can be further subdivided into dilatation-dominated transformations and shear-dominated transformations. The BCT → DC transformation in tin occurring in a diffusionless manner is an example of a dilatation-dominated transformation.

The shear dominated transformations can have a wide range of shear displacements, starting from displacements comparable in magnitude to the mean vibrational amplitude to displacements of much larger magnitude, up to 20% of the interatomic distance. Those transformations which are characterized by small shear displacements are termed *quasimartensitic*, whereas those with large shear displacements are termed truly *martensitic*.

According to this classification, a martensitic transformation is *defined* as a lattice-distortive, virtually diffusionless structural change having a dominant shear strain and associated shape change such that the strain energy dominates the kinetics and morphology during the transformation.

11.3 MORPHOLOGICAL CHARACTERISTICS

Morphology refers to the structure and form of a material. In martensitic terminology, the word is used to define the shape of a martensitic particle or the geometrical arrangement of neighbouring particles. There are two main morphological types:

1 the lath martensite, and
2 the plate martensite.

Other terms such as massive martensite and acicular martensite are best avoided. There are also other less-frequently occurring morphologies such as chevron martensite and surface martensite.

The Lath Martensite

A lath has the shape of a strip, where the thickness is less than the breadth, which in turn is less than the length. The typical thickness (the smallest dimension) of a martensitic lath is about 0.1–0.2 μm, a size barely resolvable under the optical microscope. The breadth (the intermediate dimension) of a lath is larger by about an order of magnitude than the thickness. The length (the largest dimension) of a lath is limited by the austenite grain boundary and is about 30–40 μm for a medium grain size.

Early observations on the lath product revealed only the nearly-straight boundaries of a group of laths and this led to the term massive martensite. Later experiments showed that the massive block or packet actually consisted of a number of individual laths arranged in a parallel fashion. Several such groups can form within the parent austenitic grain, the units of a group being non-parallel to the units of an adjacent group. The laths do not cross the austenite grain boundaries. Adjacent laths may be separated by high- or low-angle boundaries or may be twin related. As the laths of a group are parallel, they tend to fill the austenitic grain more or less completely, Fig. 11.4(a).

The parallel formation of laths of a group and the specific angular relationships between adjacent groups indicate a well-defined habit plane for the lath martensite. The determination of the habit plane is inherently difficult, as the laths rarely form in isolation and are also too small in size. Available experimental data on the habit plane indicate it to be $\{111\}$.

A network of tangled dislocations is the characteristic substructure of a lath, Fig. 11.4(b). The dislocation density inside a lath is usually high, in the range of 10^{15}–10^{16} m^{-2}. Cell formation within the dislocation tangles is often evident.

The lath martensite forms at higher transformation temperatures and in alloys with lower alloy content in the base metal (iron) as compared to plate martensite. The following binary systems form lath martensite: Fe–C, Fe–N, Fe–Co, Fe–Cr, Fe–Cu, Fe–Ir, Fe–Mn, Fe–Mo, Fe–Ni, Fe–Pt, Fe–Ru, Fe–Sn, Fe–V and Fe–W. In addition, a number of ternary and other multi-component systems such as Fe–Mn–C, Fe–Ni–C, Fe–Ni–Co, Fe–Ni–Mn, Fe–Ni–Ti, Fe–Ni-Cr–C, Fe–Ni–Mn–C and Fe–Ni–Co–Cr–Mn–C also exhibit lath morphology.

The Plate Martensite

The shape of a martensitic particle is often that of a disc, plate or

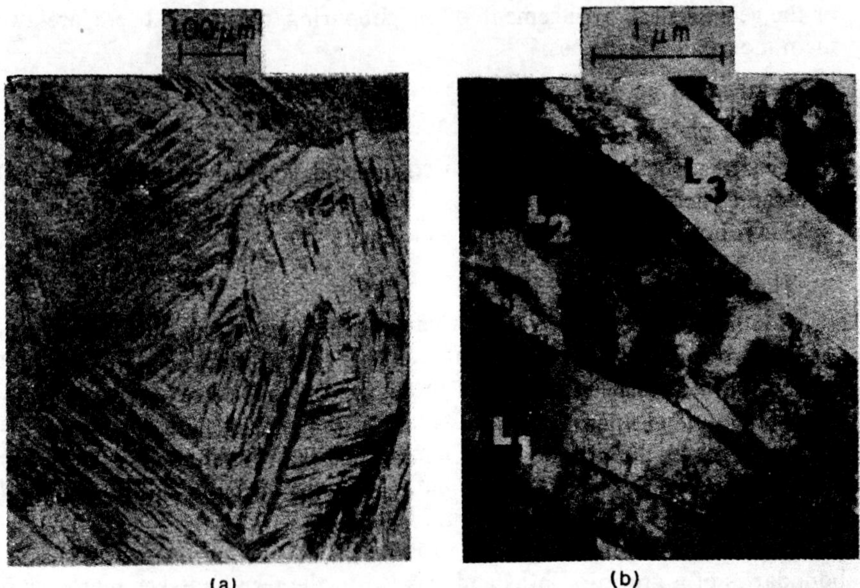

(a) (b)

Fig. 11.4 (a) The lath morphology of martensite, (b) The substructure of dislocation tangles L_1, L_2 and L_3 are individual laths.
(*Courtesy*: The Metallurgical Society of AIME, Warrendale).

biconvex lens. The term lenticular martensite also describes this shape correctly, even though it is now generally accepted that this morphology be called plate martensite.

Neighbouring plates of martensite in an austenite grain are non-parallel to one another, Fig. 11.5(a). In some cases, the ends of two neighbouring plates may be joined together. First forming plates usually extend up to the austenite grain boundaries, except in very large grains where they may not extend up to the boundary. The non-parallel formation of neighbouring plates partitions the austenite grains into smaller pockets. The plates that form subsequently are limited in size by the volume of the pockets in which they form. Even with several sub-partitioning of the austenite, the grain does not get completely filled with martensite. The transformation proceeds by the nucleation of new plates rather than by the growth of the existing plates. A cluster of plates may form more or less simultaneously in a grain and the transformation may then proceed by spreading to the adjacent grains.

Internal twins are the major substructural feature of a martensitic plate, Fig. 11.5(b). As we will see later, the internal twins arise from the second shear of the transformation. In a number of cases, the plate may have a midrib. In cases, where the plates are partly twinned, the outer rim of the plates contains a high density of dislocations. The austenite-martensite interface is not always smooth. This is especially so in cases where there is partial twinning. The plates that form at very low temperatures tend to be thin and fully twinned with smooth boundaries.

As compared to the lath martensite, the plate morphology is found gene-

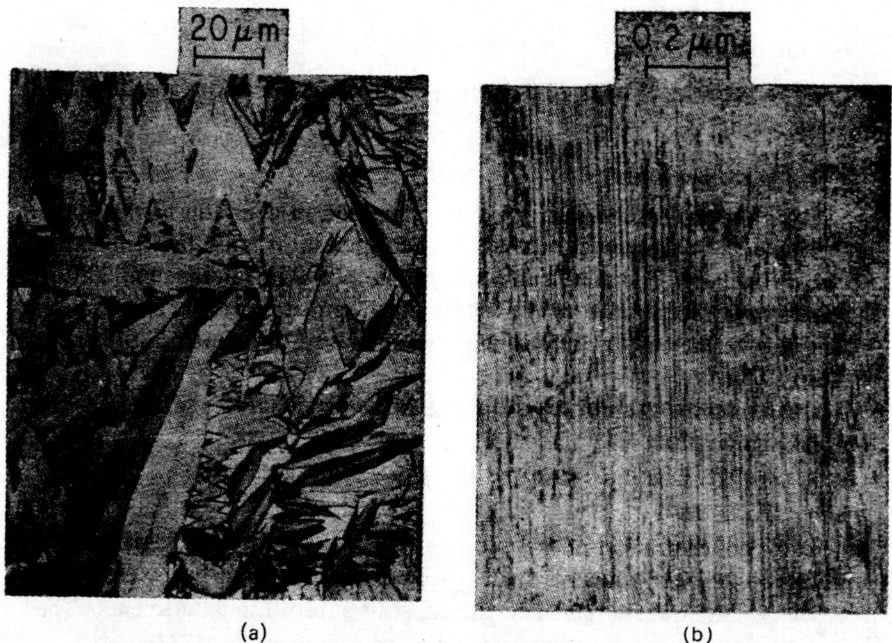

Fig. 11.5 (a) The plate morphology in a partially transformed Fe–Ni alloy. (b) The substructure of fine internal twins. .
(*Courtesy*: The Mettallurgical Society of AIME, Warrendale).

rally at lower transformation temperatures and in alloys with a higher concentration of alloying elements. Binary systems forming plate morphologies include Fe–C, Fe–N, Fe–Mn, Fe–Ni and Fe–Pt. In addition, plate martensite forms in a number of multicomponent systems such as Fe–Cr–C, Fe–Mn–C, Fe–Ni–C, Fe–Ni–Cr and Fe–Ni–Mn.

Transition in Morphology

We have already noted that lath martensite forms at higher temperatures as compared to plate martensite. Most alloys which transform at higher temperatures have also a lower alloy content. This interdependence makes it difficult to decide whether the transformation temperature determines the morphology or the alloy content. Some isothermal studies, where the transformation temperature can be varied for the same composition, indicate that the transformation temperature is the controlling factor in determining the morphology. The yield strength and the elastic modulus of both austenite and martensite, which vary with temperature and composition, may also play a role in determining the morphology.

11.4 KINETIC CHARACTERISTICS

Martensite is known to form in three different kinetic modes:

1 the athermal mode

2 the isothermal mode
3 the burst mode.

The characteristics of each of these modes of transformation are described below.

The Athermal Kinetics

Here, thermal activation plays no apparent role in the formation of martensite. The amount of martensite is a function of only the temperature to which the alloy is cooled and not the time of holding at that temperature. The athermal reaction starts at a well-defined temperature M_s. The amount of martensite increases with decreasing temperature below M_s, as illustrated in Fig. 11.2(a).

The M_s temperature of low alloy steels can be described by the following empirical equation:

$$M_s(^\circ\text{C}) = 561 - 474\ \text{C} - 33\ \text{Mn} - 17\ \text{Ni} - 17\ \text{Cr} - 21\ \text{Mo} \quad (11.1)$$

where the concentration of the alloying elements is in wt. %. In general, the effect of interstitial elements such as carbon and nitrogen in lowering M_s is an order of magnitude larger than the effect of substitutional elements. The chemical free energy change Δg for the transformation at M_s varies in magnitude from 100 to 300 MJ m^{-3}.

The progress of the athermal reaction below M_s has been described by empirical relationships such as

$$f = 1 - \exp(-1.10 \times 10^{-2} \Delta T) \quad (11.2)$$

where f is the martensitic fraction and ΔT is the degree of supercooling below M_s. Also, if we ignore the first few percent of transformation below M_s as well as the later stages, where the martensite content exceeds 60–70%, the martensitic fraction varies *linearly* with temperature. In a number of nickel and chromium steels, the slope of the linear region is directly proportional to the rate at which the free energy difference between austenite and martensite varies with temperature.

The Isothermal Kinetics

At subzero temperatures, some alloys are known to transform to martensite isothermally, i.e., as a function of time at a constant temperature, see Fig. 11.2(b). Here, the bulk transformation characteristics are measured in the absence of surface martensite. It is seen that the martensitic reaction starts slowly, then accelerates due to autocatalysis and finally decays. The transformation kinetics exhibit a typical C-curve behaviour as a function of temperature, the nose of the curve being at —140°C for an Fe–Ni–Mn alloy as shown in Fig. 11.6. In contrast to this fully isothermal behaviour, some alloys transform isothermally, only after some martensite has formed on cooling. The isothermal mode affords an opportunity to compute the activation energy for the transformation, as f can be measured as a function of time at different temperatures.

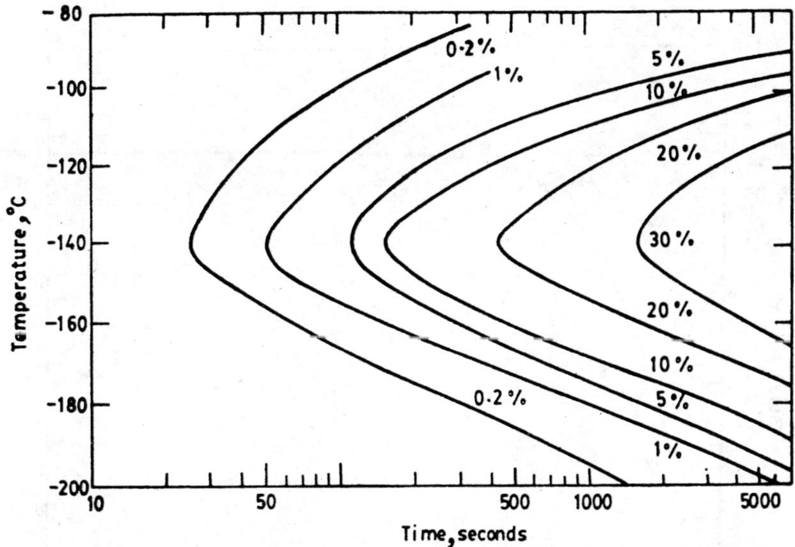

Fig. 11.6 The C-curve kinetics of the isothermal martensitic transformation in an Fe–Ni–Mn alloy.

The Burst Transformation

The martensitic transformation can set in abruptly in the form of a burst on reaching a characteristic temperature M_b, which is usually below room temperature. All the plates in a burst form in a very small fraction of a second often accompanied by an audible click, indicative of an extreme form of autocatalysis. The burst can result in considerable adiabatic heating of the specimen, which may inhibit further transformation. The transformation mode subsequent to the burst can be either athermal or isothermal. The size of a burst varies widely in different alloys from a few percent of martensite to more than fifty percent.

Factors Influencing the Kinetics

A number of variables such as the austenitic grain size, the austenitizing time and temperature, prior plastic deformation of the austenite and superimposed magnetic and elastic stress fields influence the kinetics of martensitic transformations.

Increasing the austenitic grain size increases both M_s and M_b. If the transformation is isothermal, an increase in the grain size increases the rate of isothermal transformation at a given holding temperature. Figure 11.7 shows the increase in M_b in an Fe–Ni–C alloy as a function of the austenitic grain size. The M_b temperature increases with increasing grain size up to 150 μm. Thereafter, it remains constant with further increase in grain size. As the size of the initial plates of martensite is limited by the austenite grain boundaries, fewer nucleation events are sufficient in larger grains to yield a given fraction of martensite.

The effect of the austenitizing conditions is difficult to separate from the effect of grain size and in some cases, from the effect of carbide dissolution

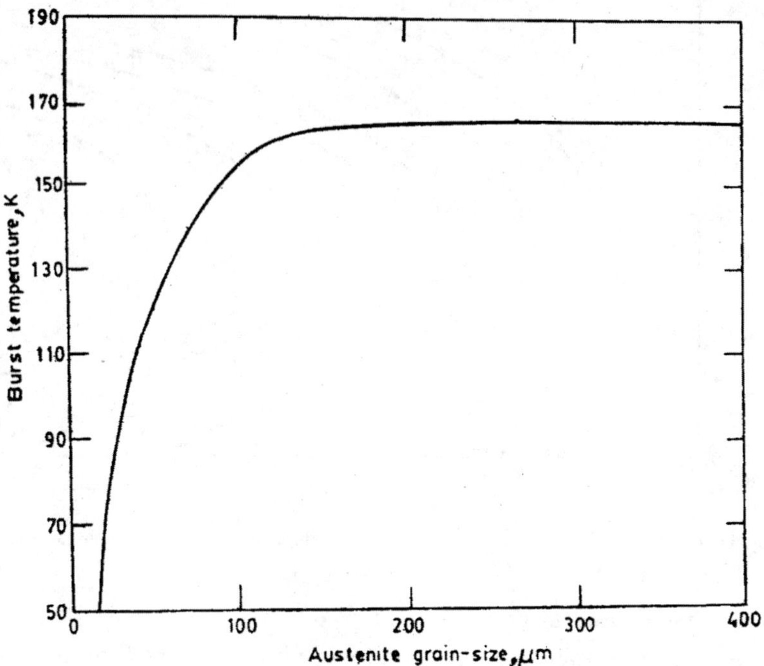

Fig. 11.7 The effect of austenitic grain size on the burst temperature.

during austenitizing. In order to overcome this difficulty, double austenitizing treatments have been carried out. After a higher-temperature anneal, the alloy is held at different lower temperatures before quenching. Such a procedure indicates that the lower holding temperature does not significantly alter the transformation kinetics.

A uniaxial stress superimposed during the transformation increases M_s, M_b or the rate of isothermal transformation, as the case may be. When the superimposed stress is less than the yield stress of austenite, the elastic stress assists the chemical driving force in inducing the transformation. This happens at temperatures slightly above M_s. At higher temperatures, the austenite deforms plastically, before the applied stress reaches a value necessary to initiate the stress-assisted transformation. Here, the plastic strain-embryos that may form in the deformed austenite can assist the transformation process. Alternatively, in some alloys the transformation is inhibited due to the work hardening of the austenite.

11.5 MODELS FOR ISOTHERMAL TRANSFORMATION KINETICS

When martensite forms isothermally, the plates grow to their full size in much less than a second. So, the transformation, spread over several hours, is controlled essentially by the nucleation rate and the mean volume of the martensitic plate per nucleation event.

Isothermal Nucleation Rate

The nucleation rate at a constant temperature is a function of the number of nucleation sites in the austenite, which is determined by the following factors:

1 the initial nucleation sites which pre-exist in the austenite
2 the newly created sites that result from autocatalysis
3 the sites that are consumed by activation into martensitic plates
4 the sites that are swept out in the volume of the transformed material.

Combining the above factors, the number of nucleation sites n_f as a function of martensitic fraction f can be written as

$$n_f = (n_i + pf - N_v)(1 - f) \tag{11.3}$$

where n_i is the number of the initial nucleation sites, p is the autocatalytic factor and N_v is the number of martensitic plates that have formed. The factor $(1 - f)$ in Eq. 11.3 accounts for the sweeping out effect.

The rate of nucleation of martensite $\dot{N}$ is given by

$$\dot{N} = n_f \nu \exp\left(-\frac{\Delta W_a}{kT}\right) \tag{11.4}$$

where ν is an appropriate vibration frequency and ΔW_a is the activation energy for nucleation.

Direct measurements of nucleation rates during isothermal martensitic transformations can be carried out by upquenching samples transformed for different lengths of time at a subzero temperature for metallographic examination. Each nucleation event at the isothermal holding temperature is a fully grown plate in the microstructure. The nucleation rate $\dot{N}$ per unit volume of austenite is derived from the following expression:

$$\dot{N} = \frac{1}{(1 - f)} \frac{dN_v}{dt} \tag{11.5}$$

N_v, the number of martensitic plates per unit volume, is evaluated from the Fullman equation applicable to thin circular discs:

$$N_v = \frac{8\bar{E}N_A}{\pi^2} \tag{11.6}$$

where $\bar{E}$ and N_A are the mean of the reciprocals of the plate lengths as intersected by a random test plane and the number of plates per unit area of a random section, respectively. Figure 11.8 shows a plot of N_v as a function of the isothermal holding time. For small martensitic fractions, the slope of this curve at any point yields directly the nucleation rate at that time. It is seen that the initial nucleation rate increases rapidly due to the autocatalytic effect.

The Mean Volume of Martensitic Plates

The mean volume of martensitic plates decreases as the martensitic fraction

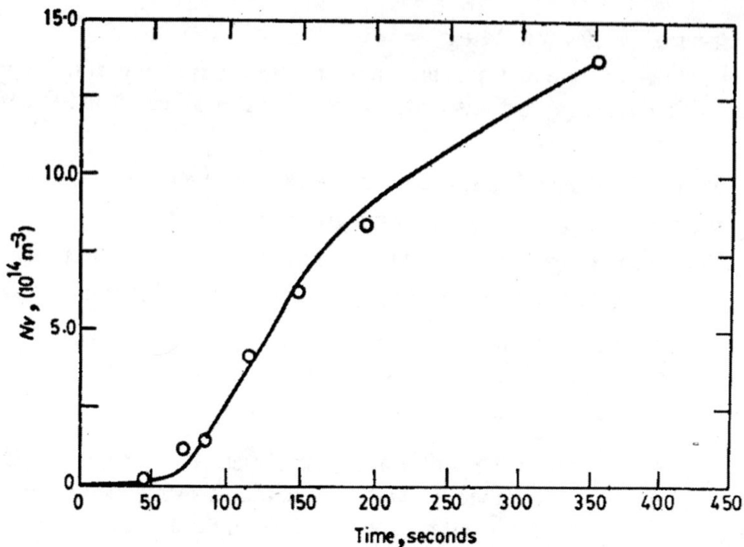

Fig. 11.8 The number of martensitic plates N_v per unit volume of austenite as a function of isothermal transformation time.

increases. The parent austenite grains get subdivided into smaller and smaller pockets as the transformation progresses, as illustrated schematically in Fig. 11.9. At any stage of the transformation, the plates are not uniformly distributed among the different grains. Due to the strong autocatalytic effect, plates tend to form in clusters around the first-nucleated plates. The clusters then spread through the parent phase to other untransformed grains.

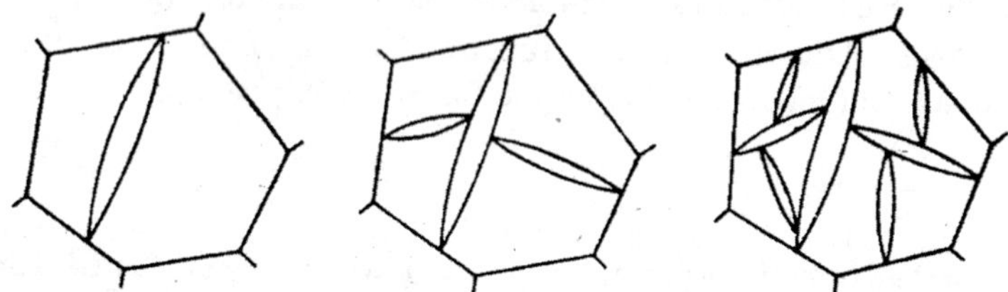

Fig. 11.9 The successive partitioning of an austenite grain by martensitic plates.

The mean volume of the plates $\bar{v}$ is directly related to N_v:

$$\bar{v} = \frac{f}{N_v} \tag{11.7}$$

f can be measured readily by point counting and N_v can be evaluated from Eq. 11.6. In Fe–Ni–Mn alloys that transform to martensite isothermally at subzero temperatures, the mean volume $\bar{v}$ decreases linearly with increasing martensitic fraction at any one test temperature.

Computation of the Isothermal Transformation Curves

The isothermal transformation rate df/dt can be written as a product of the nucleation rate $\dot{N}$ and the fraction transformed v per nucleation event, which is the instantaneous volume of a martensitic plate forming at martensitic fraction f:

$$\frac{df}{dt} = \dot{N}v \tag{11.8}$$

Note that the high velocity growth step does not come into the picture here. The instantaneous volume v is related to the mean volume $\bar{v}$ through the equation:

$$v = \bar{v} + \frac{d\bar{v}}{d \ln N_v} \tag{11.9}$$

By combining Eqs. 11.3, 11.4 and 11.7 to 11.9, an expression for the transformation kinetics is obtained:

$$\frac{df}{dt} = \left\{n_i + f\left(p - \frac{1}{\bar{v}}\right)\right\}(1 - f)\nu \exp\left(-\frac{\Delta W_a}{RT}\right)\left(\bar{v} + \frac{d\bar{v}}{d \ln N_v}\right) \tag{11.10}$$

By integrating Eq. 11.10 numerically on a digital computer, the fraction transformed f as a function of time t can be obtained. The parameters p and ΔW_a are obtained by curve-fitting the earliest stages of the transformation curve. As an example, the computed and the experimental transformation curves are compared at various isothermal test temperatures in Fig. 11.10.

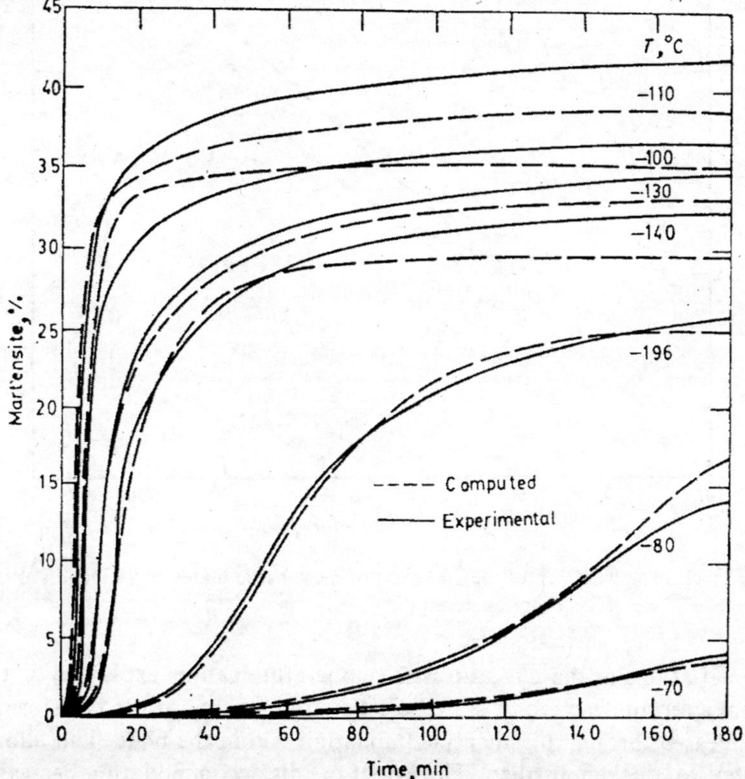

Fig. 11.10 The experimental and the computed isothermal transformation curves as a function of test temperature in an Fe–Ni–Mn alloy.

11.6 THE DIMENSIONS OF MARTENSITIC PLATES

In recent years, a number of measurements have been carried out of the mean dimensions of the martensitic plates by using quantitative metallographic methods. The results pertain to the mean semithickness $\bar{c}$, the mean radius $\bar{r}$ and the mean semithickness-to-radius ratio $\overline{c/r}$. These measurements have been made on plates formed athermally, isothermally and through bursts.

In Fe–Ni–Mn alloys, at a constant isothermal test temperature, both $\bar{c}$ and $\bar{r}$ decrease with increasing fraction of martensite. This can be attributed to the partitioning effect. The $\overline{c/r}$ ratio, however, increases with increasing martensitic fraction. The value of this ratio at $f \sim 0$ decreases nonlinearly from 0.085 at 200 K to 0.042 at 77 K, Fig. 11.11. In Fe–Ni alloys, the $\overline{c/r}$ ratio decreases linearly from 0.11 at 240 K to 0.04 at 77 K, see Fig. 11.11. When an elastic stress is superimposed, the $\overline{c/r}$ ratio increases. When the parent austenite is given a prior plastic strain, the $\overline{c/r}$ ratio decreases.

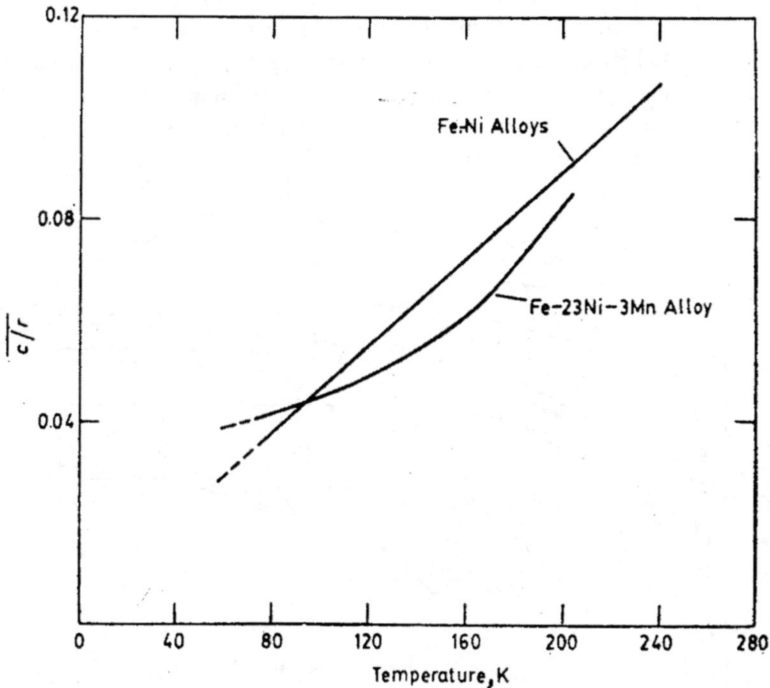

Fig. 11.11 The variation of the $\overline{c/r}$ ratio of martensitic plates as a function of transformation temperature.

The variations of the c/r ratio with temperature can be explained on the basis that a certain fraction of stored elastic strain energy of the transformation is released through plastic accommodation within the plate. This allows the plates to thicken further. The extent of this accommodation decreases with decreasing transformation temperature.

11.7 CRYSTALLOGRAPHIC FEATURES

Experimental Observations

As already pointed out, surface relief effects are observed during martensitic transformations. Parallel reference scratches on the prepolished surface of austenite get displaced in a characteristic fashion. The displaced part of the scratches remain parallel, straight and continuous across the martensite-austenite interface. This indicates that vectors in austenite transform to other vectors in martensite and planes are transformed to other planes. These are the characteristics of a homogeneous deformation. In a macroscopic sense then, the martensitic transformation can be described as a homogeneous deformation. Also, the continuity of the fiducial lines across the interface indicates that the habit plane remains unrotated.

Martensitic transformations proceed at great velocities at temperatures approaching absolute zero. This will be possible only if there is a perfect coupling between austenite and martensite, as the plate grows. For this, the austenite-martensite interface should be a glissile (coherent or semicoherent) interface (and not an incoherent boundary). It should be also invariant, so that distortions do not build up as the interface moves during growth.

The habit plane can be determined from the orientation of the austenite and the trace of the interface on two mutually-perpendicular surfaces. Such observations show that the habit plane is generally an irrational plane, as listed for a few alloys in Table 11.1. There is usually some scatter in the experimental values about those shown in the table. The orientation between austenite and martensite is determined by the back reflection Laue technique. Some experimental values are listed in Table 11.1. In the Kurdjumov-Sachs relationship, the close packed planes and directions in austenite become the close packed planes and directions in martensite.

TABLE 11.1

Crystallographic Relationships

Source	*Orientation*	*Habit plane*	*Alloy*
Kurdjumov-Sachs	$(111)_A \parallel (011)_M$ $[\bar{1}01]_A \parallel [\bar{1}\bar{1}\bar{1}]_M$	$\{225\}_A$	Fe–1.4 wt.% C
Nishiyama	$(111)_A \parallel (011)_M$ $[\bar{1}\bar{1}2]_A \parallel [0\bar{1}1]_M$	$\{259\}_A$	Fe–30 wt.% Ni
Greninger-Troiano	$(111)_A \parallel (011)_M$ within 1° $[\bar{1}01]_A \parallel [\bar{1}\bar{1}1]_M$ 5° apart	$\{3, 10, 15\}_A$	Fe–22 wt.% Ni–0.8 wt.% C

The Lattice Correspondence

In Fig. 11.12, a body centred tetragonal (BCT) unit cell is outlined within two adjacent face centred cubic (FCC) unit cells. During the transformation of austenite (FCC) to martensite (BCT) in steels, the vertical c-axis of the outlined BCT cell (which is the same as the cube edge) is presumed to contract by about 20% and the horizontal a_1 and a_2 axes (which are half of the face diagonal of the cube) to expand by 12% each, so that the BCT cell

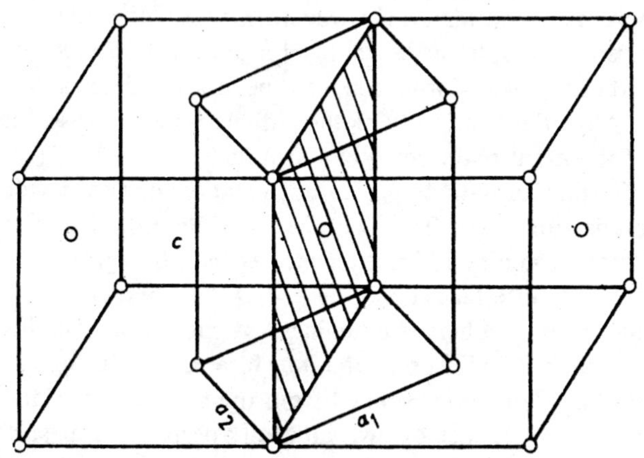

Fig. 11.12 The lattice correspondence in the FCC → BCT martensitic transformation. The face centred atoms on the front and back faces of the FCC unit cells are not shown.

acquires the cell dimensions characteristic of the product martensitic phase. Even though there is no experimental method to verify that the transformation actually takes place in this manner, these displacements correspond to easily visualized atomic movements. We can deduce readily the relationships between the planes and the directions in the two lattices. For example, in the above case, the corresponding directions are:

$$[001]_A \rightarrow [001]_M$$

$$[110]_A \rightarrow [100]_M$$

$$[\bar{1}10]_A \rightarrow [010]_M$$

where the subscripts A and M stand for austenite and martensite, respectively. A similar correspondence between crystal planes in the two lattices can also be inferred from Fig. 11.12. Such a correspondence between planes and directions of the two lattices is termed *the lattice correspondence* of the transformation.

Once the lattice correspondence is established, the lattice parameters of the two phases enable us to determine the magnitude of the atomic displacements necessary to bring about the transformation. These displacements are usually known as the Bain strain, named after E.C. Bain who first suggested the nature of the atomic displacements operative during martensitic transformations in steels. In the above case, the Bain strain expressed as strains along the principal axes of the BCT cell are:

$$\epsilon_1 = \epsilon_2 = \frac{a_M - a_A/\sqrt{2}}{a_A/\sqrt{2}} = +12\% \tag{11.11}$$

$$\epsilon_3 = \frac{c_M - a_A}{a_A} = -20\% \tag{11.12}$$

where a and c are the edge dimensions of the unit cells. The Bain strain gives the minimum atomic displacements among all conceivable correspondences between the two lattices. Even though the atomic displacements are only small fractions of the interatomic distance, the resulting strains are quite large.

Need for a Lattice Invariant Shear

Greninger and Troiano measured the shape deformation of the martensite transformation on a free surface. They attempted to explain the shape deformation as a simple shear parallel to the interface. Such a shear would have left the interface invariant. However, this simple shear failed to convert the austenite lattice to the martensite lattice. This led to the concept of a second shear which is not parallel to the habit plane. This second shear occurs on such a fine scale that it goes undetected in the shape measurements but it keeps the habit plane invariant in a macroscopic sense. The idea of a second shear has remained central to the later-developed mathematical theories.

Let a unit sphere represent the initial lattice, which is converted by the Bain strain to an ellipsoid representing the final lattice. The sphere should be strained by +12% along X_1 and X_2 axes and by −20% along the X_3 axis. A little reflection will show that the Bain strain does not leave any plane of the sphere undistorted. If an undistorted plane is to result from a homogeneous deformation, one of the strains must be positive, one must be negative and the third must be zero. In Fig. 11.13, the strain along X_3 is negative, it is positive along X_2, and zero along X_1. The planes *AOC* and *BOD* of the sphere (shown hatched in Fig. 11.13) are not distorted during the deformation, but are rotated relative to their initial positions. The purpose of the lattice invariant shear is to nullify the strain along X_1 produced by the Bain strain. The compensating strain is either slip or twinning deformation, so that the product lattice produced by the Bain strain remains unaltered.

Theories of Martensite Crystallography

The earliest theories of the crystallography of martensitic transformations are those due to Wechsler, Lieberman and Read (WLR) and Bowles and Mackenzie (BM). The elements of these theories, which are essentially equivalent, are the following:

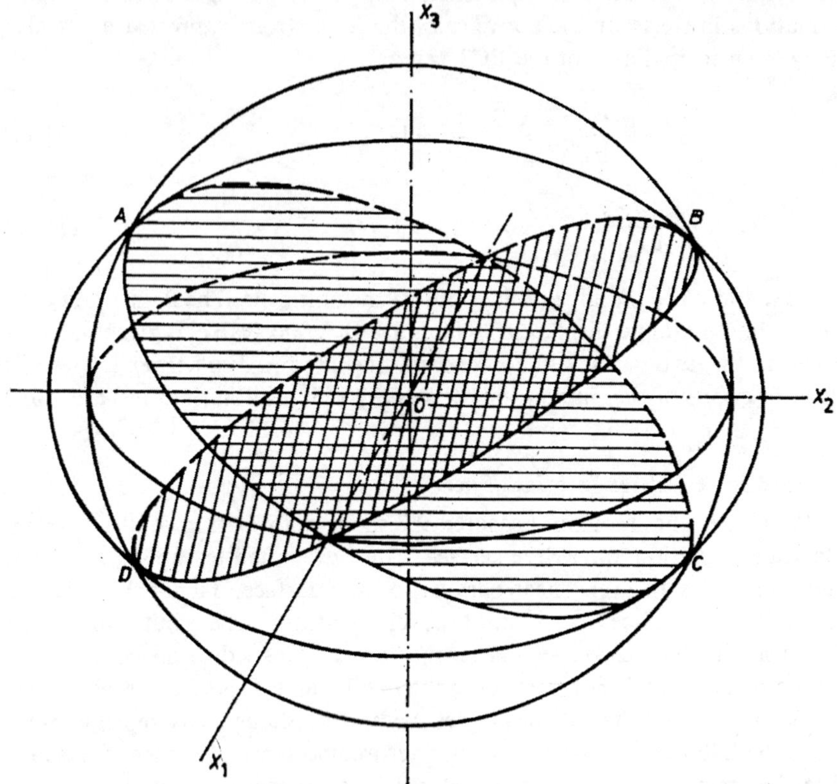

Fig. 11.13 A sphere, converted to an ellipsoid through a negative strain along X_3, a positive strain along X_2 and zero strain along X_1, leaves planes AOC and BOD undistorted but rotated.

1 The Bain strain B, with the minimum atomic displacements, converts the austenite lattice into martensite lattice.

2 A second shear called the lattice invariant shear P ensures that the distortions on the habit plane produced by the Bain strain, are removed in a macroscopic sense without changing the lattice. This shear introduces fine twins in the transformation product or slip dislocations at the interface, as illustrated in Fig. 11.14.

3 The rotation effect of the Bain strain on the habit plane is nullified by the third operation, which is a rigid body rotation R.

These three operations together constitute the invariant plane strain S:

$$S = RBP \tag{11.13}$$

The most successful aspect of these theories is that they predict the {3, 10, 15} habit plane experimentally found in Fe–Ni–C alloys, see Table 11.1. Also, the transmission electron microscopic observations of the substructure of martensite have confirmed the presence of fine internal twins in the

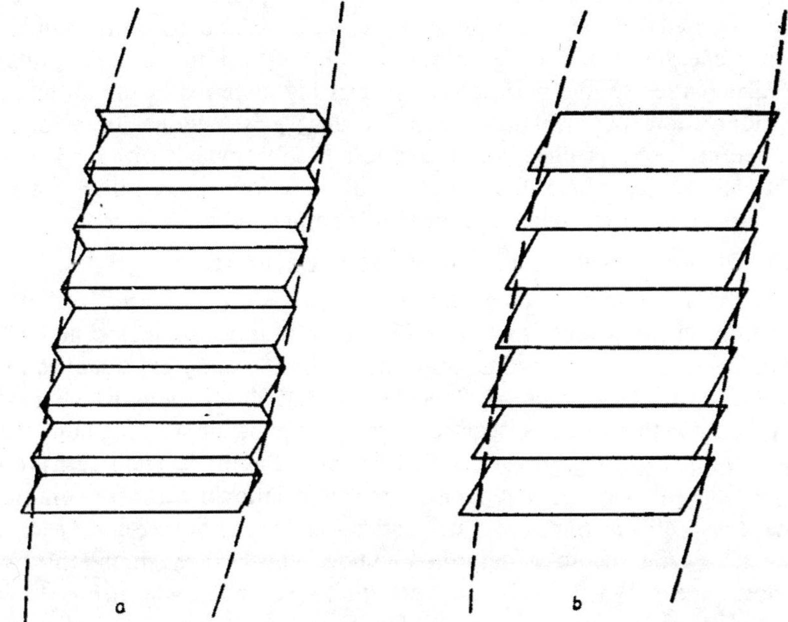

Fig. 11.14 The lattice invariant deformation by (a) twinning, or (b) slip.

martensite plate. The {259} habit is close to {3, 10, 15}, even though it is significantly different from it. The {225} habit is quite different from the above habits and is not predicted by these theories.

Recent experiments have shown the occurrence of two (or more) independent sets of twins in the same martensitic plate. The twinning planes are different in the two cases, but the twinning direction is the same. Based on such observations, the original theory can be modified to the following description:

$$S = RBP_1P_2 \tag{11.14}$$

where P_1 and P_2 replace the single lattice invariant shear P in Eq. 11.13.

11.8 NUCLEATION AND GROWTH MODELS

Difficulties with the Classical Nucleation Theory

The nucleation barrier for a phase transformation that is accompanied by appreciable strain and surface energies can be estimated from Eq. 4.31. Taking $\sigma = 0.2$ J m^{-2}, $A = 1600$ MJ m^{-3}, and $\Delta g = 200$ MJ m^{-3} at M_s, the activation energy for nucleation comes out to be $\sim$ 1000 eV. As this is prohibitively high, any model based on homogeneous nucleation is clearly ruled out. Experimental results indicate that nucleation occurs heterogeneously. With the constraints such as an invariant plane strain and a glissile interface, nucleation at the austenite grain boundaries or at the interface of inclusions as discussed in Chapter 4 appears unlikely. The strain fields of isolated dislocations in the austenite matrix cannot assist in providing the strain energies of the shape deformation and of the interface dislocations. This dilemma led to the postulate of preexisting embryos in the austenite, even though no experimental evidence for their existence has been found. Until recently, the problem of nucleation has been dealt with only in an operational sense, where the question of how the martensitic nucleus attained the size corresponding to the saddle point is left unanswered.

The Olson-Cohen Mechanism of Martensitic Nucleation

In this model, the fundamental nucleating defect is an array of lattice dislocations in the parent phase near a grain boundary, an inclusion interface or some other defect. The dislocations in the array are visualized to dissociate into two arrays of Shockley partial dislocations which repel each other, but this will be restricted, if the energy of the stacking fault that is generated is high. If, however, the energy of the faulted region turns out to be less than that of the parent phase, the dissociation process can continue spontaneously. If one Shockley partial advances for every three close-packed planes, the FCC structure is deformed homogeneously through one-third of a twinning shear. When this is combined with a second homogeneous shear, a BCC structure with the correct stacking is generated. It has been estimated that 4–5 lattice dislocations with the spacing of three close-packed planes between neighbouring dislocations will spontaneously dissociate at M_s with the available chemical driving force. A smaller array will dissociate at a higher driving force or a lower temperature, thus accounting for a spectrum of embryos of varying potency.

In this model, the isothermal nucleation is identified with the thermally-activated motion of the dislocations bounding the fault embryo. The activation energy for nucleation Q has been shown to be

$$Q = Q_0 + (\tau_a + \rho E/b + 2\sigma/nb)v^* + (\rho v^*/b)\Delta G \tag{11.15}$$

where Q_0 is the activation energy for dislocation motion in the absence of a driving force, τ_a is the athermal stress for dislocation motion, ρ is the density of atoms on the close-packed planes, b is the Burgers vector of the partial dislocations bounding the fault embryo, E is the coherency strain energy, σ is the surface energy of the austenite-martensite interface, n is the number of atom planes defining the embryo thickness, v^* is the activation volume and ΔG is the chemical free energy change per mole. Using the temperature-dependent v^* values obtained from elastic stress experiments, it has been shown that the experimental and the calculated activation energies for isothermal nucleation agree well as shown in Fig. 11.15.

The Growth Kinetics

As already pointed out, the growth kinetics of a martensitic particle cannot be described as a thermally activated atom transfer across the interface. Rather, the growth process is to be visualized as the outward motion of the interface dislocations and the nucleation of new dislocation loops at the

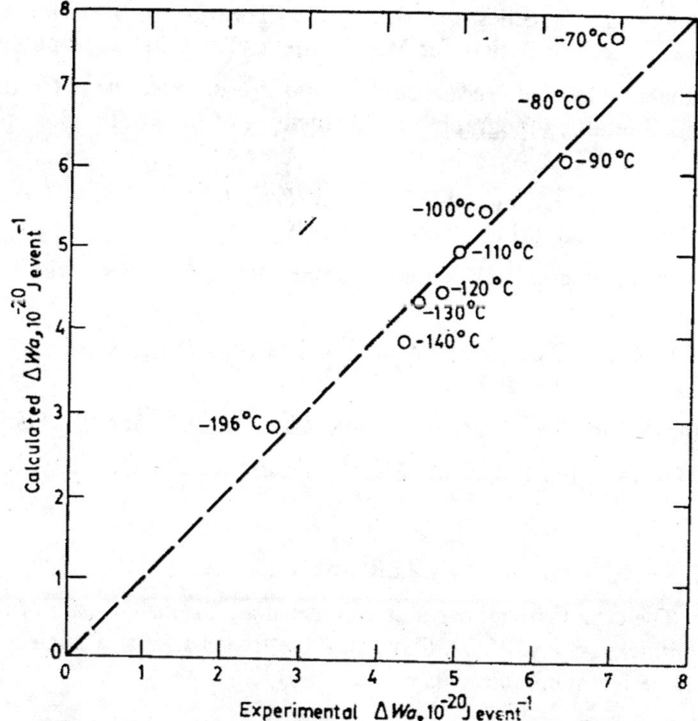

Fig. 11.15 The experimental activation energies for isothermal nucleation compared with those calculated from the Olson-Cohen model.

tips of the growing particle. A transformational stress can be calculated from the free energy gradient in the radial or thickness direction. This stress acts on the interface dislocations, just as a mechanical stress acts on the lattice dislocations during plastic deformation. The velocity of the interfacial dislocations v_d under the transformational stress τ_{th} in the thickening direction has been computed using Gilman's relationship:

$$v_d = v_{\lim} \exp\left(-\frac{\mathscr{D}}{\tau_{th}}\right) \tag{11.16}$$

where $v_{\lim}$ is the limiting velocity and $\mathscr{D}$ is the drag coefficient. From the calculated velocity, the time for the growth of a plate to full size, after the first few steps of the growth start-up have occurred, can be computed. This time turns out to be a few microseconds, agreeing with the experimental observation that martensitic plates grow at very high velocities.

FURTHER READING

M. Cohen and C.M. Wayman, Fundamentals of Martensitic Reactions, in *Metallurgical Treatises*, The Metallurgical Society of AIME, Warrendale, PA, p. 445 (1981).

V. Raghavan, Kinetics of Martensitic Transformations, in *Cohen Symposium*, American Society for Metals, Metals Park, OH, to be published.

D.S. Lieberman, *The Mechanism of Phase Transformations in Crystalline Solids*, Inst. Metals Monograph No. 33, Institute of Metals, London, p. 167 (1969).

M. Cohen, G.B. Olson and P.C. Clapp, *Proc. Int. Conf. on Martensitic Transformations*, Cambridge, MA, p. 1 (1979).

A.B. Greninger and A.R. Troiano, *Trans. Amer. Inst. Min. Met. Engrs.*, **185**, 590 (1949).

M.S. Wechsler, D.S. Lieberman and T.A. Read, *Trans. Amer. Inst. Min. Met. Engrs.*, **197**, 1503 (1953).

J.S. Bowles and J.K. Mackenzie, *Acta Metall.*, **2**, 129 and 224 (1954).

G.B. Olson and M. Cohen, *Metall. Trans. A.*, **7A**, 1897, 1905, 1915 (1976).

EXERCISES

11.1 The strain energy per unit volume of a martensitic plate of radius r and semithickness c is $A(c/r)$. Show that the free energy of a nucleus of a given volume is a minimum, when

$$c = \sqrt{\frac{\sigma r}{A}},$$

where σ is the interfacial energy.

11.2 Prove that the instantaneous volume of a martensitic plate v is related to the mean volume $\bar{v}$ through the relation

$$v = \bar{v} + \frac{\mathrm{d}\bar{v}}{\mathrm{d}\ln N_v}$$

where N_v is the number of martensitic plates per unit volume of austenite.

11.3 Show that, for a martensitic plate that is in thermoelastic equilibrium, the following condition must be met:

$$\Delta g + 2A(c/r) = 0$$

Assume that the stress required to move the interfacial dislocations during growth is negligible.

11.4 In the analysis of the austenite partitioning by Fisher, Hollomon and Turnbull, after N_v plates have formed, the austenite is divided into pockets of $q/(qN_v + 1)$ in size, where q is the average grain volume. It is assumed that each plate transforms a constant fraction m of a pocket. Then

$$\frac{\mathrm{d}f}{\mathrm{d}N_v} = \frac{mq(1-f)}{(qN_v + 1)}$$

Show that

(a) $\dfrac{df}{dN_v} = mq(1 - f)^{(m+1)/m}$

(b) $N_v = -\dfrac{1}{q}\{1 - (1 - f)^{-1/m}\}$

(c) Compute the number of martensitic plates at $f = 0.01$, 0.1, 0.5 and 0.7. Assume $q = 10^{-13}$ m^3 and $m = 0.05$. Comment on the results you obtain.

12

Spinodal Decomposition

In the introductory chapter, a distinction was made between transformations that occur homogeneously and those that occur heterogeneously through nucleation and growth. Spinodal decomposition belongs to the former category. It starts with small compositional fluctuations spread out over the whole volume undergoing transformation. The fluctuations grow in intensity as a function of time, eventually yielding the final product phases. The transformation mechanism is basically different from the nucleation-and-growth type of reactions discussed up to now. Spinodal decomposition proceeds with a progressive decrease in free energy from the very beginning, without any need for mounting a nucleation barrier.

12.1 THE SPINODAL CURVE

Figure 12.1(a) shows schematically the free-energy/composition relationships for a binary system exhibiting a miscibility gap. x and y are the two inflection points on this curve. The free energy of a mixture of two compositions c_1 and c_2 that lie *outside* the inflection points is at d corresponding to the mean composition c_0 on the straight line ab. The free energy of this mixture is *higher* than that of the homogeneous solution c_0. The free energy of a mixture of two compositions c_1' and c_2 that lie *inside* the inflection points is at d' on the straight line $a'b'$. The free energy of the mixture here is *lower* than that of the uniform composition c_0'.

A spinodal decomposition brought about by compositional fluctuations can take place spontaneously with a decrease in free energy, if the overall composition lies within the two inflection points. For all compositions that lie between the inflection points, the second derivative of free energy with respect to composition is negative. When the compositions corresponding to the inflection points are plotted as a function of temperature (on a phase diagram), we obtain the *chemical spinodal curve*, Fig. 12.1(b).

1 Along the spinodal, $\dfrac{d^2G}{dc^2} = 0$

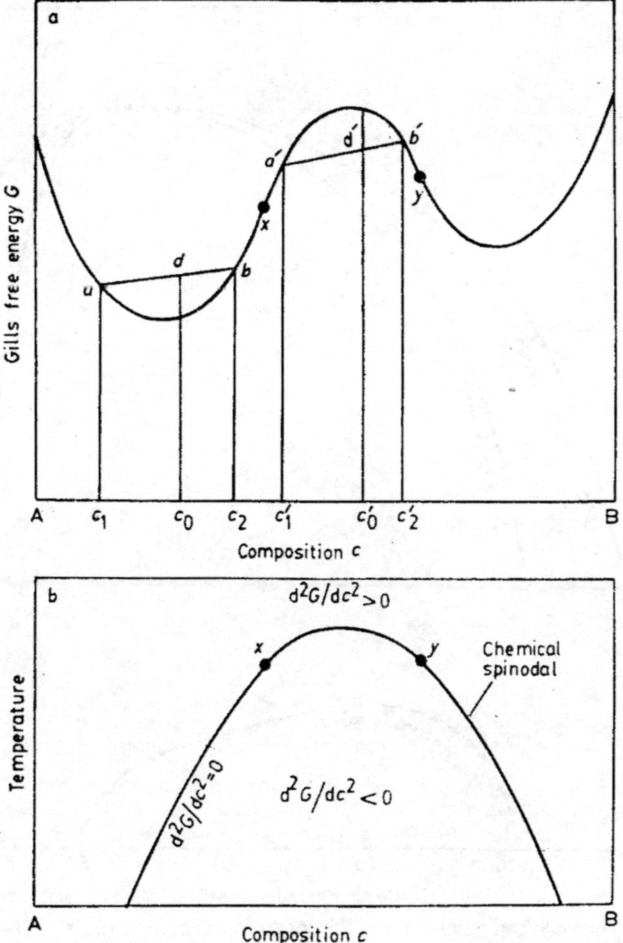

Fig. 12.1 (a) The free energy-composition curve for a binary system exhibiting a miscibility gap. (b) The chemical spinodal.

2 Inside the spinodal, $\dfrac{d^2G}{dc^2} < 0$

3 Outside the spinodal, $\dfrac{d^2G}{dc^2} > 0$.

Figure 12.2 shows the miscibility gap and the chemical spinodal for the Au–Ni system. Note that, at the critical temperature T_c where the miscibility gap disappears, the phase boundary and the chemical spinodal are tangential to each other.

12.2 FREE ENERGY OF COMPOSITIONAL FLUCTUATIONS

In Chapter 3, page 43, we considered the free energy change during a nucleation-and-growth type of transformation with a compositional change.

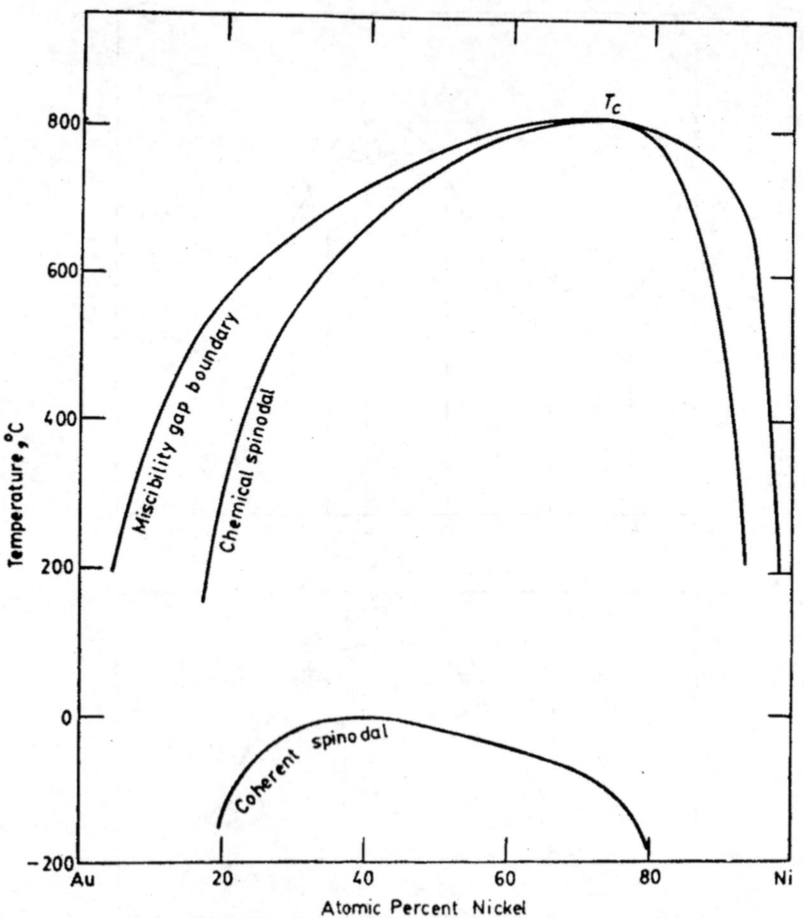

Fig. 12.2 The chemical spinodal and the coherent spinodal in the Au–Ni system.

We now address the case of compositional change in which nucleation and growth are not operative.

The free energy of a solid solution of nonuniform composition can be divided into two terms:

1 The free energy of an infinitesimal volume of composition c integrated over the whole volume under consideration

2 The free energy associated with the compositional gradient across any infinitesimal volume, again integrated over the whole volume.

The Gradient Energy

The free energy due to a composition gradient is analogous to the usual surface energy term. If, for example, there is a coherent interface between two phases (α_1 and α_2) of different compositions, surface energy is associated with this interface. The energy is proportional to the square of the composition difference across the interface. Similarly, when compositional fluctuations occur, the bond distribution changes and the corresponding energy change in any infinitesimal volume is proportional to the square of the concentration gradient across the volume. If this is integrated over the total volume of the system, we have a gradient energy term, which can be considered as the surface energy of the diffuse interface. The constant of proportionality is called the gradient energy coefficient κ. It is approximately equal to the product of the enthalpy of mixing per unit volume at $c = 0.5$ and the square of the interatomic distance. The enthalpy of mixing for systems exhibiting a miscibility gap is positive, recall the interaction energy V for such systems from Chapter 3. A typical values of κ is 10^{-10} J m^{-1}.

Helmholtz Free Energy of Sinusoidal Fluctuations

Consider for the sake of simplicity a one-dimensional compositional fluctuation along the length x of a bar of overall composition $\bar{c}$ and of constant cross-sectional area A'. The Helmholtz free energy F of the bar is given by

$$F = A'\int_0^x f(c)\,\mathrm{d}x + A'\int_0^x \kappa\left(\frac{\mathrm{d}c}{\mathrm{d}x}\right)^2 \mathrm{d}x \tag{12.1}$$

where $f(c)$ is the free energy per unit volume of composition c.

If the fluctuation is sinusoidal, the composition c at position x can be written as

$$c - \bar{c} = A\cos kx \tag{12.2}$$

where A is the amplitude of the fluctuation and k is the wave number, related to the wavelength λ through $k = 2\pi/\lambda$. The concentration gradient is obtained by differentiating Eq. 12.2 with respect to x:

$$\frac{\mathrm{d}c}{\mathrm{d}x} = kA\sin kx \tag{12.3}$$

Expanding $f(c)$ about the mean composition $\bar{c}$ and neglecting terms higher than second order, we have

$$f(c) = f(\bar{c}) + (c - \bar{c})\left(\frac{\mathrm{d}f}{\mathrm{d}c}\right)_{c=\bar{c}} + \frac{1}{2}(c - \bar{c})^2\left(\frac{\mathrm{d}^2f}{\mathrm{d}c^2}\right)_{c=\bar{c}} \tag{12.4}$$

Substituting Eqs. 12.2 to 12.4 in 12.1, we obtain

$$F = A'\int_0^x\left[f(\bar{c}) + A\cos kx\left(\frac{\mathrm{d}f}{\mathrm{d}c}\right)_{c=\bar{c}} + \frac{1}{2}A^2\cos^2 kx\left(\frac{\mathrm{d}^2f}{\mathrm{d}c^2}\right)_{c=\bar{c}}\right]\mathrm{d}x$$

$$+ A'\int_0^x \kappa k^2A^2\sin^2 kx\,\mathrm{d}x$$

Using the relationships

$$\cos^2 kx = \frac{1 + \cos 2kx}{2}$$

and

$$\sin^2 kx = \frac{1 - \cos 2kx}{2}$$

we obtain

$$F = A'x\left[f(\bar{c}) + \frac{A^2}{4}\left(\frac{d^2f}{dc^2}\right)_{c=\bar{c}} + \frac{\kappa k^2 A^2}{2}\right]$$
$$+ A'\left[A\left(\frac{df}{dc}\right)_{c=\bar{c}}\frac{\sin kx}{k} + \frac{A^2}{4}\left(\frac{d^2f}{dc^2}\right)_{c=\bar{c}}\frac{\sin 2kx}{2k} - \frac{\kappa k^2 A^2}{2}\frac{\sin 2kx}{2k}\right] \quad (12.5)$$

Conservation of mass during the fluctuation in a solution of initial composition $\bar{c}$ requires

$$\int_0^x (c - \bar{c})\, dx = 0 \quad (12.6)$$

Using Eq. 12.2 in 12.6, we have the condition for conservation as

$$\sin kx = 0 \quad (12.7)$$

So, the terms within the last square brackets on the right side of Eq. 12.5 are zero. Noting that $A'x$ is the volume of the bar, the free energy per unit volume is

$$f = f(\bar{c}) + \frac{A^2}{4}\left(\frac{d^2f}{dc^2}\right)_{c=\bar{c}} + \frac{\kappa k^2 A^2}{2} \quad (12.8)$$

Therefore, the free energy change due to the sinusoidal compositional fluctuation is

$$\Delta f = \frac{A^2}{4}\left[\left(\frac{d^2f}{dc^2}\right)_{c=\bar{c}} + 2\kappa k^2\right] \quad (12.9)$$

It is easy to see from Eq. 12.9 that, on the spinodal curve where $d^2f/dc^2 = 0$, $\Delta f = 0$ only when $k = 0$ (i.e., $\lambda = \infty$). For a fluctuation of finite wavelength, Δf can equal zero only *within* the spinodal. For a given composition and temperature within the spinodal, there is a critical k_c (and λ_c) at which $\Delta f = 0$. We can write

1 $\Delta f = 0$ when $\lambda = \lambda_c$
2 $\Delta f < 0$ when $\lambda > \lambda_c$
3 $\Delta f > 0$ when $\lambda < \lambda_c$.

Only wavelengths longer than the critical can form spontaneously. The critical condition is obtained by setting the quantity within the square brackets of Eq. 12.9 equal to zero:

$$\lambda_c = \left[-\frac{8\pi^2\kappa}{\left(\frac{d^2f}{dc^2}\right)_{c=\bar{c}}}\right]^{1/2} \quad (12.10)$$

The inclusion of the gradient energy term is an important step in our understanding of spinodal decomposition. If the classical diffusion equation is used within the spinodal for describing the kinetics without the gradient energy term, the fastest growing wavelength will be the smallest possible one. This will then predict the separation of the two phases with a spacing between them equal to the lattice parameter. This is neither realistic nor in agreement with the experimentally observed spacing of 50–100 Å between the phases.

Coherency Strains

Another improvement made by Cahn was to take into account the strain energy associated with the compositional fluctuation. Inasmuch as coherency is maintained during the fluctuation, elastic strains will be introduced as the lattice parameter of the solid solution varies with composition. Following Eq. 4.19, the misfit due to coherency is given by

$$\delta = \frac{a - a_0}{a_0} \tag{12.11}$$

where a_0 is the unstrained lattice parameter of the initial composition $\bar{c}$ and a is the strained parameter due to the fluctuation. The strain η per unit change in composition is taken to be constant (Vegard's law):

$$\eta = \frac{1}{a_0}\frac{da}{dc} \tag{12.12}$$

Combining Eqs. 12.11 and 12.12, we obtain

$$\delta = \eta(c - \bar{c}) \tag{12.13}$$

The total strain energy ϵ^T in the bar of constant cross-section A' and length x due to a unidirectional fluctuation along x is given by (recall Eq. 4.21),

$$\epsilon^T = A' \int_0^x \frac{Y\delta^2}{1 - \nu}\, dx \tag{12.14}$$

where Y is the Young's modulus and ν is the Poisson's ratio of the solid solution. Taking Y and ν to be independent of composition and using Eqs. 12.13 and 12.2, we can write

$$\epsilon^T = \frac{A' Y \eta^2 A^2}{2(1 - \nu)} \int_0^x (1 + \cos 2kx)\, dx \tag{12.15}$$

This yields the strain energy per unit volume as

$$\epsilon = \frac{Y\eta^2 A^2}{2(1 - \nu)} \tag{12.16}$$

Equation 12.9 for the free energy change can now be modified to take the strain energy into account:

$$\Delta f = \frac{A^2}{4}\left[\left(\frac{d^2 f}{dc^2}\right)_{c=\bar{c}} + \frac{2\eta^2 Y}{(1-\nu)} + 2\kappa k^2\right] \tag{12.17}$$

The free energy change will now become zero only when

$$\left(\frac{d^2 f}{dc^2}\right)_{c=\bar{c}} + \frac{2\eta^2 Y}{(1-\nu)} = 0 \tag{12.18}$$

Corresponding to this condition, a *coherent spinodal* can be defined which lies inside the chemical spinodal, as illustrated in Fig. 12.2 for the Au–Ni system.

If the crystal is anisotropic, the strain energy will vary as a function of crystal direction. Compositional fluctuations along those crystal directions, which minimize the strain energy, are energetically favoured. Such fluctuations outgrow fluctuations along other directions, the result being that the final product phases lie along *specific* crystallographic directions.

12.3 THE KINETICS OF SPINODAL DECOMPOSITION

The compositional fluctuations during spinodal decomposition are brought about by long range diffusion. As the solid solution is initially homogeneous with an uniform composition of $\bar{c}$, the diffusion is uphill here, in contrast to the usual downhill diffusion through the matrix during growth of a precipitate particle, recall Section 5.2. Therefore, here the chemical potential gradients are used in place of the concentration gradients in the equation for the diffusional flux.

The Diffusion Equation

The flux J under a chemical potential gradient along the x direction is given by:

$$J = -M\frac{\partial(\mu_B - \mu_A)}{\partial x} \tag{12.19}$$

where M is the *mobility* and μ_A and μ_B are the *chemical potentials* of the A and B components in the binary solution. The quantity $(\mu_B - \mu_A)$ is the change in the free energy, when a unit amount of B atoms is added and a unit amount of A atoms is removed from a large quantity of the solution in a reversible manner. For a homogeneous solution,

$$\mu_B - \mu_A = \frac{\partial f}{\partial c_B} \tag{12.20}$$

Substituting Eq. 12.20 in 12.19, we have

$$J = -M\frac{\partial}{\partial x}\left(\frac{\partial f}{\partial c_B}\right)$$

$$= -M\frac{\partial^2 f}{\partial c^2}\frac{\partial c}{\partial x} \tag{12.21}$$

By comparing Eq. 12.21 with Fick's first law (Eq. 2.1), we note

$$D = M\left(\frac{\partial^2 f}{\partial c^2}\right) \tag{12.22}$$

where D is the diffusion coefficient. As M is always positive, the diffusion coefficient takes the sign of $(\partial^2 f/\partial c^2)$:

1 outside the spinodal, $D > 0$
2 on the spinodal, $D = 0$
3 inside the spinodal, $D < 0$.

When a concentration gradient is present such that the average concentration changes appreciably within the interaction range of an atom, the gradient will change the chemical potential. If the plot of composition (concentration of B atoms) versus distance has a positive curvature at a point, a B atom at this point will sense more B neighbours in its environment than the local composition would indicate. The change in the chemical potential due to this effect turns out to be proportional to the curvature and the gradient energy coefficient κ. So, the chemical potential difference in Eq. 12.20 is modified as:

$$\mu_B - \mu_A = \frac{\partial f}{\partial c} - 2\kappa\frac{\partial^2 c}{\partial x^2} \tag{12.23}$$

Rewriting the equation for flux:

$$\begin{aligned} J &= -M\frac{\partial}{\partial x}\left(\frac{\partial f}{\partial c} - 2\kappa\frac{\partial^2 c}{\partial x^2}\right) \\ &= -M\left(\frac{\partial^2 f}{\partial c^2}\frac{\partial c}{\partial x} - 2\kappa\frac{\partial^3 c}{\partial x^3}\right) \end{aligned} \tag{12.24}$$

Under nonsteady state conditions, we can write from Fick's second law:

$$\frac{\partial c}{\partial t} = -\frac{\partial J}{\partial x} = M\left(\frac{\partial^2 f}{\partial c^2}\frac{\partial^2 c}{\partial x^2} - 2\kappa\frac{\partial^4 c}{\partial x^4}\right) \tag{12.25}$$

provided that $(\partial^2 f/\partial c^2)$ can be taken to be independent of position (and therefore of composition). This approximation is valid only during the early stages of the decomposition, when the amplitude of the fluctuation is still small. Equation 12.25 can also be modified to include the coherency strain energy.

Solution to the Diffusion Equation

A general solution to Eq. 12.25 is

$$c(x, t) - \bar{c} = A(k, t)\cos kx \tag{12.26}$$

where $A(k, t)$ is given by

$$A(k, t) = A(k, 0)\exp R(k)t \tag{12.27}$$

$A(k, 0)$ is the initial amplitude of the Fourier component of wave number k.

$R(k)$ is called the *amplification factor* and is defined by

$$R(k) = -Mk^2 \left[\left(\frac{d^2f}{dc^2} \right)_{c=\bar{c}} + 2\kappa k^2 \right] \tag{12.28}$$

For those wavelengths, where the free energy change Δf defined by Eq. 12.9 is negative, $R(k)$ is positive. Where Δf is positive, the amplification factor is negative.

By setting $dR(k)/dk = 0$, we find that the maximum in the amplification factor occurs when

$$-2Mk \left(\frac{d^2f}{dc^2} \right)_{c=\bar{c}} - 8M\kappa k^3 = 0 \tag{12.29}$$

So,

$$k_{max} = \left[-\frac{\left(\frac{d^2f}{dc^2} \right)}{4\kappa} \right]^{1/2}$$

$$= \frac{1}{\sqrt{2}} k_{\bar{c}} \tag{12.30}$$

Also,

$$\lambda_{max} = \sqrt{2}\, \lambda_c \tag{12.31}$$

λ_{max} is that wavelength that grows fastest in intensity among the compositional fluctuations.

Comparison with Experimental Results

Several experimental checks of the diffusion equation 12.25 are now available for spinodal decomposition. The early observations of satellites about the Bragg peaks in small angle x-ray scattering experiments confirm the compositional modulations during precipitation attributed to the spinodal mechanism. A fluctuation results in modulation in the lattice parameter as well as the atomic scattering factor, both of which are a function of composition. The contribution due to atomic scattering is found to be dominant for the satellites about the (000) peak, i.e., for a small Bragg angle.

Any arbitrary compositional fluctuation due to thermal energy can be expressed as the sum of a series of sinusoidal fluctuations of different A and k values. Since $R(k)$ occurs within the exponential in Eq. 12.27, the wavelength λ_{max} corresponding to R_{max} easily outgrows the other wavelengths. This accounts for the uniform spacing of precipitate particles that are believed to originate from spinodal decomposition. Moreover, this relatively uniform particle size and spacing is expected to resist coarsening tendencies.

The precipitate morphology observed in isotropic materials such as glass undergoing spinodal decomposition is a random arrangement of the phases. In contrast, crystalline systems exhibiting elastic anisotropy tend to have the precipitates aligned in rows along certain crystallographic directions.

The kinetics of spinodal decomposition have been studied using the small angle scattering technique. The log of intensity plotted against time in Fig. 12.3 shows a linear relationship, in agreement with Eq. 12.27. The slope of the lines in the figure is the amplification factor $R(k)$. For the smallest wavelength indicated ($\lambda = 29$ Å), $R(k)$ is negative, which means that this wavelength is decreasing in intensity. The critical wavelength λ_c for which $R(k) = 0$, lies between 29 and 34 Å. $\lambda_{max} = \sqrt{2}\,\lambda_c$ lies around 46 Å. Among the wavelengths shown in Fig. 12.3, the one with $\lambda = 46.6$ Å is increasing most rapidly in intensity with time.

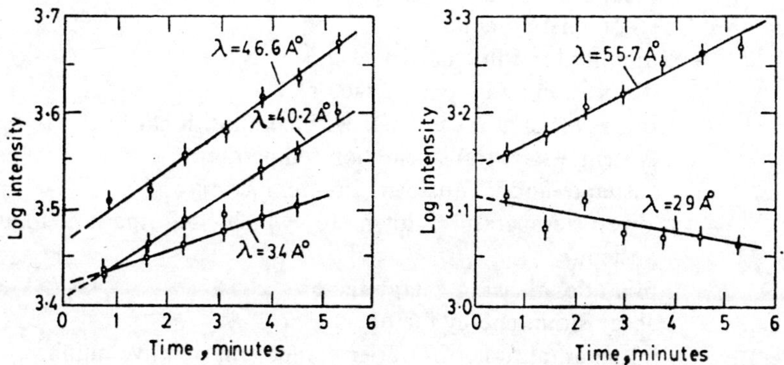

Fig. 12.3 Intensity versus time plots in small angle x-ray scattering experiments during spinodal decomposition in an Al–22 at. % Zn alloy.

FURTHER READING

J.W. Cahn, *Acta Metall.*, **9**, 795 (1961).

J.E. Hilliard, *Phase Transformations*, Amer. Soc. Metals, Metals Park, OH, p. 497 (1970).

EXERCISES

12.1 Prove the relationship given in Eq. 12.10.

12.2 Show that the fastest growing wavelength during spinodal decomposition is $\sqrt{2}$ times the critical wavelength.

12.3 Explain why a uniform size of the precipitate particles increases the resistance to coarsening.

List of Symbols

A	area, cross-sectional area
A	strain energy factor
A	amplitude of fluctuation
A'	dislocation strain energy factor
A'	cross-sectional area of bar with compositional fluctuation
A_s	austenite-start transformation temperature
A_f	austenite-finish transformation temperature
A_d	lowest temperature down to which deformation induces austenite
A_{dil}	dilatational strain energy factor
A_{shear}	shear strain energy factor
$A\ (k, 0)$	initial amplitude of Fourier component of wave number k
$A\ (k, t)$	amplitude of Fourier component of wave number k at time t
a	thickness of thin film or disc
a	unit cell axis
a	lattice parameter or atom spacing
a_0	lattice parameter
$\overset{0}{a}_m$	natural lattice constant or atom spacing of matrix
$\overset{0}{a}_p$	natural lattice constant or atom spacing of particle
a_m	actual lattice constant or atom spacing of matrix
a_p	actual lattice constant or atom spacing of particle
B	Bain strain
b	Burgers vector of dislocation
b	Burgers vector of partial dislocations bounding fault embryo
C, C_1, C_2, C_3	constants
c	composition
c	concentration of atomic specie per unit volume
c	unit cell axis
c	semiaxis of oblate or prolate spheroidal particle
c	semithickness of disc-shaped particle
c	semithickness of martensitic plate
$\bar{c}$	average concentration or composition
$\bar{c}$	initial concentration of solute in matrix
$\bar{c}$	mean semithickness of martensitic plates
$\overline{c/r}$	mean semithickness-to-radius ratio of martensitic plates
c'	composition seen by the ith particle at large distances

c' a composition in the $(\alpha + \beta)$ region
c^* critical composition in nucleation
c^* critical semithickness of disc-shaped particle
Δc concentration difference producing diffusion
c_0 average concentration
c_0 composition at which free energies of α and β phases are equal
c_{11} elastic constant
c_{12} elastic constant
c_A concentration of A component
c_B concentration of B component
c_b concentration of solute at the boundary
c_p heat capacity at constant pressure
c_s concentration at the surface
c_p^s heat capacity of solid solution at constant pressure
c_α concentration in the α matrix
$c_{\alpha\beta}$ concentration in the α matrix in equilibrium with a β particle
$c_{\beta\alpha}$ concentration in the β particle in equilibrium with the α matrix
$c_{\gamma\alpha}$ composition of γ in equilibrium with the α phase
c_{Fe_3C} composition of Fe_3C
$c_{\alpha-Fe_3C}$ composition of α in equilibrium with Fe_3C
$c_{\gamma-Fe_3C}$ composition of γ in equilibrium with Fe_3C
$c(x, t)$ concentration at position x at time t
$c_{\alpha\beta}^r$ concentration in α in equilibrium with a β particle of radius r
$c_{\alpha\beta}^{\bar{r}}$ concentration in α in equilibrium with a β particle of radius $\bar{r}$
$c_{\alpha\beta}^i$ concentration in α in equilibrium with a β particle formed initially at temperature T
$c_{\beta\alpha}^i$ concentration in β in equilibrium with α matrix, formed initially at temperature T
$c_{\alpha\beta}^\infty$ concentration in α in equilibrium with a β particle of radius ∞
$c(\infty)$ concentration in α in equilibrium with a β particle of radius ∞
D diffusion coefficient or diffusivity
D_A intrinsic diffusivity of A component
D_B intrinsic diffusivity of B component
$\tilde{D}$ chemical interdiffusion coefficient
D_0 pre-exponential constant or frequency factor
D_b boundary diffusion coefficient
D_l lattice diffusion coefficient
D_α diffusion coefficient for diffusion in α phase
D_γ diffusion coefficient for diffusion in γ phase
D_c^γ diffusion coefficient of carbon in γ phase
$\bar{D}$ mean grain diameter
$\bar{D}_0$ mean grain diameter at time $= 0$
$\bar{D}_t$ mean grain diameter at time t
$\mathcal{D}$ drag coefficient
E internal energy

$\bar{E}$	mean of the reciprocals of the plate lengths on a random test plane
e	electronic charge
erf	error function
F	Helmholtz free energy per mole
$f(c)$	Helmholtz free energy of composition c per unit volume
Δf	Helmholtz free energy change per unit volume
f	maximum dragging force of a solute on a migrating boundary
f	martensitic fraction
f	a function
f_β	volume fraction of β precipitates
G	Gibbs free energy
ΔG	Gibbs free energy change per mole
ΔG	free energy change during nucleation or transformation
ΔG_D	free energy barrier for diffusion across interface
ΔG_f	free energy of formation of vacancies
ΔG_m	free energy of motion of a diffusing specie
ΔG_{het}	free energy change during heterogeneous nucleation
ΔG_{homo}	free energy change during homogeneous nucleation
ΔG^*	critical free energy barrier for nucleation
ΔG^*_{het}	critical free energy barrier for heterogeneous nucleation
ΔG^*_{homo}	critical free energy barrier for homogeneous nucleation
ΔG^*_i	nucleation barrier for the ith type of sites
ΔG_{max}	barrier maximum during nucleation on dislocations
ΔG_{min}	metastable minimum during nucleation on dislocations
$G^\alpha_{\bar{c}}$	Gibbs free energy of α phase of composition $\bar{c}$
$G^\beta_{c_2}$	Gibbs free energy of β phase of composition c_2
g_α	Gibbs free energy of α per unit volume
g_β	Gibbs free energy of β per unit volume
Δg	Gibbs free energy change per unit volume
Δg_{net}	net Gibbs free energy change per unit volume
H	enthalpy
H_0	residual enthaply at 0 K
ΔH	enthalpy change per mole
ΔH_f	enthalpy of formation of vacancies
ΔH_m	enthalpy of motion of a diffusing specie
Δh	enthalpy change per unit volume
I	rate of nucleation per unit volume of the parent phase
I_i	initial nucleation rate
I_s	steady state nucleation rate
I_t	nucleation rate at time t
I_{het}	heterogeneous nucleation rate
I_{homo}	homogeneous nucleation rate
I_{cm}	nucleation rate of cementite per unit area of austenite grain boundary

I_f nucleation rate of ferrite per unit area of the austenite-cementite interface
I'_f nucleation rate of ferrite per cementite particle
I_p nucleation rate of pearlite per unit area of the austenite grain boundary
I'_p nucleation rate of pearlite per cementite particle
J diffusional flux
J_A diffusional flux of A atoms
J_B diffusional flux of B atoms
K reaction rate constant
k Boltzmann's constant
k diffusion distance
k wave number
k_c critical wave number
k_{max} wave number of the fastest-growing fluctuation
L_v length of grain edges per unit volume
l effective thickness of the diffusion zone
M mobility
$\overline{M}$ mean of reciprocals of the diameters of circular intersections of spherical particles with a random test plane
M_b martensite burst temperature
M_d highest temperature up to which deformation induces martensite
M_f martensite finish temperature
M_s martensite start temperature
M_s saturation magnetization
m a constant
m a constant fraction
N number of nucleation sites
N^* number of critical-sized particles
N_0 Avogadro's number
N_A number of martensitic plates per unit area of a random test plane
N_{AA} number of A-A bonds per mole
N_{AB} number of A-B bonds per mole
N_{BB} number of B-B bonds per mole
N_T total number of particles
N_i number of nuclei with the nucleation barrier of ΔG_i^*
N_p number of particles
N_v number of nucleating particles per unit volume
N_v number of grain corners per unit volume
N_v number of precipitate particles per unit volume
N_v number of martensitic plates per unit volume
n time exponent
n number of moles of diffusing specie

n	number of diffusing defects per unit volume
n	number of atoms
n	number of atom planes defining the embryo thickness in the Olson-Cohen model
n_i	number of initial nucleation sites in martensitic transformation
n_f	number of nucleation sites at martensitic fraction f
P	pressure
P	probability that the first ferrite nucleus has formed on the austenite-cementite interface
P	lattice invariant shear
P_0	equilibrium pressure
P_c	critical transition pressure
$P_{A(B)}$	probability that a B atom has a A neighbour
$P_{B(A)}$	probability that a A atom has a B neighbour
p	autocatalytic factor
Q	activation energy for diffusion
Q	activation energy for martensite nucleation in the Olson-Cohen model
Q_0	activation energy for martensite nucleation in the absence of driving force in the Olson-Cohen model
q	average grain volume
R	gas constant
R	radius of curvature of spherical cap
R	radius of spherical grain or radius of curvature of curved grain boundary
R	rigid body rotation
R^*	critical radius of curvature of spherical cap
$R(k)$	amplification factor of fluctuation of wave number k
R_{max}	amplification factor of fastest-growing fluctuation
r	radius of spherical particle
r	semiaxis of oblate or prolate spheroidal particle
r	radius of disc-shaped particle
r	radius of circle of intersection
r	radius of cylindrical particle
r	atomic radius
$\bar{r}$	average radius of precipitate particles
$\bar{r}$	mean radius of martensitic plates
r^*	radius of critical-sized particle
r_1	radius of precipitate particle 1
r_2	radius of precipitate particle 2
r_i	radius of the ith precipitate particle
r_i	initial radius
r_m	radius of fastest growing particle
r_p	radius of precipitate particle
$\bar{r}_1$	average radius of particles at time t_1
$\bar{r}_2$	average radius of particles at time t_2

$\bar{r}_3$	average radius of particles at time t_3
$\bar{r}^0$	average radius of particles initially
$\bar{r}(t)$	average radius of particles at time t
r_m^0	radius of fastest growing particles at time = 0
S	entropy per mole
S	invariant plane strain
ΔS	entropy change per mole
S_A	surface area of particle
S_0	interlamellar spacing
S_c	configurational entropy
S_v	grain boundary area per unit volume
S_{min}	minimum interlamellar spacing
$S_v(0)$	interfacial area per unit volume at time = 0
$S_v(t)$	interfacial area per unit volume at time t
s	extent of movement of markers
s	number of interfacial atoms
s^*	number of atoms facing critical particle of radius r^*
T	temperature
T_c	critical temperature
T_c	Curie temperature
T_0	equilibrium temperature
T_m	melting point in kelvin
ΔT	degree of supercooling
$T_{I\max}$	temperature of maximum rate of nucleation
t	time
t_f	time for completion of transformation
U	linear growth rate
U_x	linear growth rate in x direction
U_y	linear growth rate in y direction
U_z	linear growth rate in z direction
U_i	linear growth rate of ith particle
$U(0)$	interfacial energy per unit volume at time = 0
$U(t)$	interfacial energy per unit volume at time t
V	volume of particle
V	molar volume
V	binding energy of solute to the boundary
$\bar{V}$	interaction energy
V^*	critical volume of nucleating particle
ΔV	volume change during transformation
V_0	total volume of the system
V_β	volume of β particle
V_i	volume of ith particle
V_{AA}	energy of an A-A bond
V_{AB}	energy of an A-B bond
V_{BB}	energy of a B-B bond
v	instantaneous volume of martensitic plate

v	velocity of markers
v	volume per atom
$\bar{v}$	mean volume of martensitic plates
v^*	activation volume
v_d	dislocation velocity
v_{lim}	limiting dislocation velocity
ΔW_a	activation energy for martensitic nucleation
X	volume fraction transformed
X	fraction recrystallized
X	fraction of recovery
X	fractional decrease of interfacial energy
X	volume fraction of precipitate particles
X_1, X_2, X_3	axis of sphere-ellipsoid transformation by Bain strain
X_A	mole fraction of A atoms
X_B	mole fraction of B atoms
X_{ext}	extended volume fraction transformed (not corrected for impingement)
x	diffusion direction or distance
x	length of bar undergoing fluctuations in composition
Y	Young's modulus
Y_m	Young's modulus of matrix
Y_p	Young's modulus of particle
y	ratio of particle radii, r/r_i
y	distance
Z	coordination number
z	valence
z	a variable
α	a phase, solid solution or matrix
α	number of moles of diffusing specie per unit area in thin film
α	clustering or short range ordering parameter
α	a constant
α	a geometrical factor
β	a phase
β	a particle
γ	a phase
Γ	number of solute atoms per unit area of boundary
Γ	a constant
δ	a phase
δ	spacing between adjacent planes
δ	a distance parameter
δ	actual or optimum misfit
δ^0	natural misfit
ϵ	strain energy per unit volume of particle
ϵ_p	strain energy per particle
ϵ_m	coherency strain in matrix
ϵ_p	coherency strain in particle

ϵ	coherency strain in both particle and matrix
$\epsilon_1, \epsilon_2, \epsilon_3$	principal strains in Bajn deformation
ϵ^T	total coherency strain energy in bar with compositional fluctuation
η	strain per unit change in composition
η	an integration variable
η	a variable
θ	a phase
θ	contact angle
2θ	included lens angle
θ'	a transition phase
θ''	a transition phase
κ	gradient energy coefficient
κ	a constant
λ	wavelength
λ	a variable
λ_c	critical wavelength
λ_{max}	wavelength of fastest growing fluctuation
λ_j	atomic jump distance across an interface
μ	shear modulus
μ_m	shear modulus of matrix
μ_A	chemical potential of component A
μ_B	chemical potential of component B
ν	lattice vibration frequency
ν'	jump frequency of a diffusing atom
ν	Poisson's ratio
ν_m	Poisson's ratio of matrix
ν_p	Poisson's ratio of particle
ρ	density of atoms on close packed planes
σ	surface energy of an interface per unit area
σ	electrical conductivity
σ	flow stress of partially recovered solid
σ_m	flow stress of deformed material
σ_0	flow stress of undeformed material
$\sigma_{\alpha\alpha}$	surface energy of α-α boundary per unit area
$\sigma_{\alpha\beta}$	surface energy of α-β boundary per unit area
$\sigma_{\beta\gamma}$	surface energy of β-γ boundary per unit area
τ	time
τ	delay time
τ_a	athermal stress to move a dislocation
τ_{th}	transformational stress in the thickening direction
ϕ	shear angle of transformation
ϕ	function
ψ	angle
ω	time constant
ω	frequency factor

Index